HEART
OF THE
BOMB

•

The Dangerous Allure
of Weapons Work

•
•
•

DEBRA ROSENTHAL

▲
▼▼

Addison-Wesley Publishing Company, Inc.

Reading, Massachusetts   Menlo Park, California   New York
Don Mills, Ontario   Wokingham, England   Amsterdam   Bonn
Sydney   Singapore   Tokyo   Madrid   San Juan

Library of Congress Cataloging-in-Publication Data

Rosenthal, Debra.
    At the heart of the bomb : the dangerous allure of weapons work / Debra
    Rosenthal.
        p.   cm.
    Includes bibliographical references.
    ISBN 0-201-19794-4
    1. Nuclear weapons—Moral and ethical aspects.   2. Nuclear
engineers—United States—Attitudes.   3. Nuclear weapons—United
States—Moral and ethical aspects.   4. Sandia National Laboratories—
Officials and employees—Attitudes.   5. Los Alamos National
Laboratory—Officials and employees—Attitudes.   6. Nuclear
engineers—New Mexico—Attitudes.
U264.3.R68   1990
172'.422—dc20   89-27187

Copyright © 1990 by Debra Rosenthal

Cover design by Amy Bernstein
Text design by Anna George
Set in 10-point Caledonia by G & S Typesetters, Inc., Austin, TX

ABCDEFGHIJ-MW-9543210

First printing, May 1990

For John Manley

CONTENTS

ACKNOWLEDGMENTS

This book is based on over 260 hours of interviews with eighty-five people who work or have worked in the two nuclear weapons design facilities in New Mexico. Forty-one of those people were from Los Alamos National Laboratory, and the remainder were from Sandia. (A few had worked at both weapons laboratories.)

Thirty-six people called me to volunteer after reading in their laboratories' newsletters that I wanted to interview scientists and engineers about their social responsibility. Some said they volunteered because they thought themselves unlike their colleagues, others because they thought themselves perfectly typical, and a few simply because they wanted a chance to talk about their lives and work. Based on their suggestions, I contacted others at the laboratories who seemed likely to cover the spectrum of religious and political views, as well as more easily quantified things such as age, race, and work history. Only three declined my request for an interview. Although this was not a random sample, I am confident that the people described in this book represent the range of opinion within the weapons laboratories, if not the precise distribution.

There is no clear consensus on the politics and morality of the national defense mission at the weapons laboratories, although both laboratories harbor persistent rumors to the contrary. I began conducting interviews for this book in 1984, and while the baseline of my impressions was established at that time, I have kept in touch with people at both laboratories up to the present. It is indicative of the insular and timeless nature of weapons work that I see no sign that recent changes in domestic and international politics have had any serious or lasting effect on the attitudes and ideas of people in the laboratories. For example, the new found warmth between the United States and the Soviet Union seems to have met the same mixed reviews from weapons researchers that greeted former President Reagan's most strident anti-Soviet statements.

However, like other Department of Energy facilities, the weapons laboratories have been hit hard by recent charges of lax security. Some of the attitudes and behaviors people told me about in 1985 might seem like more serious indiscretions in the current climate of suspicion, and I suspect many would speak less freely today. To protect their identities, each has been given a false first name, and I have taken the precaution of changing minor biographical details, like adding or subtracting a few years from someone's age. Any annoying ambiguity about the specifics of someone's job is due either to my fear that saying too much might make them too easily identifiable, or to their own vagueness. I also interviewed a dozen people in opposition to the weapons laboratories, most but not all of whom would call themselves peace activists. With their permission, some people are called by their true first and last names.

I am grateful to everyone who took the time to offer information and advice for this project. I owe a special debt to my husband and family for their support and to Marilyn Gayle Hoff and John Manley, whose fine minds and goodwill were especially important to me.

For over a year I had a weekly luncheon date with a diverse

group of people concerned about the nuclear arms situation. I always learned something, and thank the participants for sharing their perspectives. Discussions with Harjit Ahluwahlia, Roger Anderson, Barbara Francis, Junella Haynes, David Morrissey, Philip Roeder, Wilhelm Rosenblatt, Peggy Rosenthal, Bob Russell, and Mark Rutledge were also very helpful to me.

I would also like to thank Vince Ercolano, Ronna Kalish, Bruce Martin, Ron Noggle, and Maryellen Wolfe for research assistance and Dee Bauer, Sue Fitzmaurice, Patty Gegick, and Robin Jones for secretarial help. I alone am responsible for my errors.

I began background research for this project as a participant in a seminar sponsored by the National Endowment for the Humanities in which I learned a great deal about applied ethics from the participants and from James Childress, the seminar director. Scientific research can be very costly, as scientists from the weapons labs will attest. The research behind this book was not, with much of the expense of traveling and buying coffee for the people I interviewed covered by a small grant from the University of New Mexico.

The people I interviewed for this book opened their homes and entrusted the stories of their lives to a stranger. My greatest debt is to them.

1

THE LAND OF
ENCHANTMENT

.
.
.

New Mexico encourages magical and mystical thinking. You start in the east where New Mexico borders Texas, and drive into a stiff wind on Interstate 40 toward a sunset and across a desert right out of a classic western movie. Only daydreams enliven the flat miles of untidy scrub brush tenuously rooted in orange dirt that sustains mainly jackrabbits and clusters of doleful-looking cattle. On the shoulder of the road, handmade crosses, some wired with plastic flowers, mark the place a spirit departed. In this region where mystics of one sort or another have traveled for centuries, signs along the interstate now urge travelers to make it to "Tucumcari Tonite!" The main attraction there is cheap, clean motel rooms.

The law ends and the Wild West begins after a gradual descent from the high plains into the Pecos River Valley. Eventually the horizon becomes jagged, the road rises into foothills, and suddenly you shoot into Tijeras Canyon, which separates the Sandia Mountains running north from the Manzanos to the south. After twenty minutes on the highway between the two mountain ranges, the

1

canyon opens into Albuquerque, pretty much right in the middle of this big square state and home to nearly one-third of New Mexico's residents.

Most people keep on going through the city, past the Wyoming Street exit to Kirtland Air Force Base, down into the valley of the disappointingly thin and muddy Rio Grande. They head into more desert and badlands blackened by volcanic ash, with perhaps a stop in Gallup. Indians from the Navajo reservation go there to buy groceries, pickup trucks, and alcohol—the latter being their number one cause of death. Tourists shop for Indian rugs or turquoise and silver jewelry and then breathe a sigh of relief when they finally leave the low brown hills and head toward the Grand Canyon, or maybe California, where the greens are really green, not the dry, smoky grays of chamisa and sage.

They might get a hint of the special hidden beauty here if their timing was right, if they left Albuquerque at sunset and their eyes caught the rearview mirror. *Sandia* means "watermelon," and these ten-thousand-foot-high mountains are so named because the setting sun turns their irregular oval profile a warm, sweet red.

It is the northern part of New Mexico, though, that inspired the "Land of Enchantment" slogan on the state's license plates. Nearly every vista north of Albuquerque is painfully beautiful. The interstate highway running north from Albuquerque to Santa Fe, the state capital (with its own unfortunate slogan, "The city different"), passes through gently curving hills dotted with dark green juniper and piñon bushes at intervals so regular that it seems a draftsman, rather than the lack of water, must have dictated their placement.

Red and gold mesas appear on the horizon. Signs inform you when you enter and leave the lands belonging to the pueblo Indians who settled along the valley of the Rio Grande. Eventually conquered and at least nominally Christianized, these people still do the Corn Dance, and other traditional rituals, some secret, others at which visitors are warmly welcomed. Further north are the Sangre de Cristo Mountains, the southernmost chain in the Rockies. They also change hue in the sunset, turning the color, so it was thought, of the "Blood of Christ." In these harsh mountains, sects of renegade Catholics called Penitentes reenact Christ's pas-

sions each Easter, complete with self-flagellations and a mock cru-
cifixion. But to the west lies a completely different world, a city
called Los Alamos. Home of the atom bomb, this is where ratio-
nality, we are told, not magic or religion, rules.

J. Robert Oppenheimer suggested that the bomb be built at Los
Alamos because the place was beautiful and isolated. Today there
is a three-lane highway winding up increasingly steep cliffs and
hillsides, an improved bridge next to the old wooden Otowi Bridge
over the Rio Grande, a paved back road that curves south by the
Los Alamos ski basin and back to Albuquerque, and an airport on
the edge of town. Until 1957 the guard tower on the main road
was manned, and visitors to Los Alamos needed either identification or a damn good reason to get in. Now the guard tower is
empty. Tourist or resident or Soviet spy, you can drive right into
the town that arguably has done more to change human history
than any other place on earth.

The labs at Los Alamos were founded secretly in 1942, but
within a few years, Los Alamos scientists had established a less
isolated and more convenient outpost in Albuquerque. Sandia National Laboratories, which became independent of Los Alamos in
1949, is located inside Kirtland Air Force Base at the southeast
edge of the city. There, with more secrecy than the elves at Christ-
mastime, people do the ordinance engineering for every nuclear
weapon in the American arsenal.

A few years ago, on an airplane bound for New Mexico, I found
myself seated next to an engineer who made his living designing
nuclear weapons in the Land of Enchantment. Paul was a pleasant
and engaging conversationalist. We talked about current events,
and he came across as a perfectly agreeable and benign technocrat,
interpreting the human condition as a series of problems easily
resolved by social engineering.

Talk between strangers who know they will never meet again is
usually superficial, but it can also go the other way. After a couple
of hours of conversation, overcome with curiosity and my courage
bolstered by our relative anonymity, I asked Paul how he felt
about working on nuclear weapons.

The question at first unsettled him. He dropped his eyes and said he did not usually discuss such things because they seemed too personal. Nonetheless, buoyed perhaps by the same false intimacy that had led me to inquire, he proceeded to tell me this story.

Whenever he finished working on a bomb project, Paul said, he always felt miserable. He would think he had done the wrong thing. He would be overcome with guilt. But then, he explained, he'd begin to rationalize his involvement with the weapons.

During his career at Sandia, Paul had developed a standardized, systematic procedure for easing his conscience. First he would tell himself that if he didn't do the work, someone else would. Then he would spread the responsibility to the American public, since we elect the politicians who make the decisions to build nuclear weapons. He would remind himself that the Russians build bombs, so we have to build bombs.

None of these thoughts actually made him feel much better, he admitted, but by the time he had finished this line of reasoning, he would be so engrossed in a new project and so entranced by the technical problems it presented that he could put aside completely his moral reservations.

Paul had used the same rationalizations for nineteen years. He had no idea how other weapons scientists and engineers felt. He had never once discussed with any of his colleagues the moral dimensions of their work.

Countless novels, short stories, and plays feature the major characters who first moved to Los Alamos to support work on the atom bomb. Oppenheimer's comment that the Manhattan Project captivated scientists because the challenges of building the bomb were "technically sweet" and his quote from the Bhagavad-Gita ("I am become Death, the destroyer of worlds"), applied to the Trinity Test, are cited everywhere. The power struggle between Oppenheimer and Edward Teller, partly over the wisdom of an all-out effort to build an H-bomb, is easily cast as a Battle of Titans.

But even within the impassioned atmosphere of World War II, at the University of Chicago's Metallurgical Laboratory and in the place with the code address "P.O. Box 1663, Santa Fe," there were those who wondered if the bomb might be a mistake, a cure worse than the disease itself. Questions about the wisdom of pursuing the bomb and, later, of using it on two Japanese cities, were ultimately deflected by the participants' commitment to finishing what they had begun, by political momentum, and by the adroit use of secrecy restrictions to quash discussion among the doubters.

But what motivates contemporary nuclear weapons designers? Very soon now the cold war will have waged for half a century— a long time to adhere to orthodoxy, to continue momentum, to maintain silence. Are the scientists and engineers expanding and enhancing our nuclear arsenal today the last cold warriors, standing on the "shoulders of giants," the ultimate true believers? Or, like Paul, are they simply self-absorbed tinkerers, content so long as they are allowed to play with their sophisticated but deadly toys?

I have always thought of scientists as representing, however awkwardly, the height of Enlightenment values. They use the scientific method to dispel superstition. They do not fear creativity and can challenge even the most entrenched orthodoxy. They think logically, and they have an unrelenting passion for truth. Their knowledge allows them to manipulate and control the capricious forces of nature, thus making us all less vulnerable. So why, I wondered, did Paul sound more like a victim than a master?

2

SANDIA BOMB
& NOVELTY

.
.
.

Kirtland Air Force Base bustles during the day with military effi-
ciency. Roughly twenty thousand people work there, including the
employees of Sandia National Laboratories. In the middle of the
night it is a supernatural jungle of barracks, offices, and chain link
fences illuminated by flashing orange, red, white, and yellow
lights. During normal business hours you can get onto the base to
visit the National Atomic Museum, a cavernous building filled
with bomb and missile prototypes. Once the sun goes down, you
need a friend with a pass.

Our goal was to circle the Manzano Mountain. Kirtland Base
completely encloses the Manzano Mountain, which is northern-
most in the Manzano chain. Surrounded by four security fences,
one electrified, Manzano is a base within a base, easily visible from
the freeway through Tijeras Canyon. Its four softly rounded peaks
are unimpressive compared to the Sandias, seeming more like hills
than mountains.

At one time you could leave the freeway and travel back east a bit into Coyote Canyon where locals would practice target shooting at cardboard boxes and beer cans. Climb some pine and piñon-covered hills and you could see the convoluted Manzano Mountain and the few small buildings and trailers planted on its slopes. Coyote Canyon has since been closed off, so to get a decent look at the Manzano Base requires driving back down into the city, onto Kirtland, and through the business part of the Air Force base to where the buildings give way to fields filled with rabbits. A lot of the buildings have heroic statues at their entrances: missiles of various sorts, some nose up, others nose down.

The "World's Largest All-Wooden Structure" is on the road to the Manzano Mountain. A huge trestle, like half a bridge to nowhere, held together by dowels, it is used by the Air Force Weapons Laboratory to test the effects of the electromagnetic pulse (EMP) emitted when a nuclear bomb is exploded. Scientists and engineers from the Air Force create an artificial EMP to test its impact on equipment like airplanes. Constructed with practically no metal of any kind, the trestle does not distort the effects or their measurement. The EMP disrupts electrical and magnetic fields. In a nuclear attack, it might keep your car from starting. It certainly would make your radio and TV worthless, and make military command and control very difficult. In 1989 a Washington, D.C.–based technology watchdog group, the Foundation for Economic Trends, sought a court injunction against any further EMP tests at Kirtland Base, claiming the electromagnetic radiation may be harmful. Military officials denied immediate or chronic health threats to workers or nearby residents.[1]

At night the area next to the trestle is sometimes marked by a fairy circle of red lights on fifty-foot poles, with nothing in the middle. A riding stable and a golf course lie further down the road.

Of the nineteen major government organizations housed at Kirtland, nine are intimately tied to the development and deployment of nuclear weapons. Here the Defense Nuclear Agency's Field Command Office monitors the location, security, and safety

---

[1] See Lawrence Spohn, "The Trestle," *Albuquerque Tribune*, May 8, 1989, p. C-1; Spohn, "Tower of Power," ibid., p. A-1.

of all American nuclear weapons, between twenty and thirty thousand by most estimates. Some of these weapons are maintained by Kirtland Base's 3098th Aviation Depot Squadron. A Sandia administrator told me, matter-of-factly, that Albuquerque is home to the largest single storage site for nuclear weapons in the world. The bombs are on Manzano Base—or rather, *in* it. The four dusty gray hills of Manzano Mountain are hollow.

• • •

America's nuclear weapons are the responsibility of the Department of Energy, formerly the Atomic Energy Commission. Until recent investigations of DOE nuclear facilities splashed its name across the front pages, most Americans probably assumed the department had something to do with oil and solar power. It does, but with the emphasis always on the relationship between energy resources and national security. (When the old AEC was reorganized, the licensing and inspection of civilian nuclear power plants was handed over to a separate agency, the Nuclear Regulatory Commission.) The country's largest DOE field office, the Albuquerque Operations Office, is a few blocks from Sandia Laboratories on Kirtland Base.

The DOE coordinates the nine major facilities around the country that develop, produce, and test all nuclear weapons intended for the U.S. military. For example, the nuclear reactors in Hanford, Washington, produce plutonium and tritium, the elements that fuel the reactions in thermonuclear weapons. The Pantex plant in Amarillo, Texas, made famous by Grace Mojtabai in *Blessed Assurance*, is another of the nine. All the parts of nuclear weapons are assembled into the real thing, real warheads, in Amarillo.[2]

Scientists at Los Alamos and at Lawrence Livermore National Laboratories in California design the "guts" of nuclear warheads, more poetically called "physics packages"—the parts that blow up.

---

[2] A. G. Mojtabai, *Blessed Assurance: At Home with the Bomb in Amarillo, Texas* (Boston: Houghton Mifflin, 1986).

They figure out how to make a bomb with certain characteristics: this much explosive yield, this much radiation of such and such a sort, weighing about so many pounds and occupying some specific amount of space. To be useful these packages must be "weaponized." That work is done at Sandia National Laboratories. Engineers at Sandia design all the "peripheral parts and components" that detonate a warhead, and they figure out how to put the guts and parts into "delivery packages"—bodies that can be dropped, shot, propelled through water, or whatever the military orders.

The roughly eight thousand regular employees at each of these three laboratories are civilians, the policies that manage them and the paychecks they draw issued by civilian operators. Although most work at their home laboratories, a few people from Los Alamos are assigned to Lawrence Livermore, and Sandia has 1,100 people stationed in a separate facility near there. Another hundred or so Sandia employees stay at the Tonepah Test Range and the Nevada Test Site, where the bombs are detonated in underground tunnels to ensure that they work and to find out more about *how* they work. As Government Owned Contractor Operated laboratories (GOCOs), Los Alamos and Lawrence Livermore are run by the University of California; Sandia by AT&T (formerly Bell Laboratories). AT&T makes only a symbolic profit for its efforts, one dollar a year. The University of California, on the other hand, receives substantial direct compensation for the services it provides as manager of Los Alamos, amounting to $5.6 million in 1987–88.

Middle-aged engineers can recall asking Bell Labs recruiters "What's Sandia?", and it is still far from a household word. In contrast, protesters argue every five years that the University of California's Board of Regents should not renew its contract to operate Los Alamos and Lawrence Livermore.

The Department of Energy, designating the three laboratories as "prime contractors," provides most of their funding and sponsors most of their weapons, energy, and basic research projects. Based on funding designation, both Sandia and Los Alamos have seen an increase in the proportion of their work devoted to defense. From the end of the Carter administration to the mid-1980s the Los Alamos defense projects budget increased from just over half to 72 percent of its total expenditures. The traditionally more defense-oriented Sandia engineering laboratory in Albuquerque had only 16 percent of its budget going to nonweapons research in 1985. Sandia and Los Alamos each have budgets in the neighborhood of one billion dollars a year. The public affairs director of Sandia is fond of pointing out that about one out of every thousand dollars in the federal budget goes to his laboratory.

•   •   •

Beyond the stables and the golf course, near the very southern edge of Kirtland Base, lies another Department of Energy GOCO laboratory, the Inhalation Toxicology Research Institute. This lab studies the effects of airborne toxins, primarily radioactive, on living creatures, primarily beagles. Viewed from the lab where the dogs die, the four hills of the Manzano Mountain were merely darker shadows against the night sky, with modest red lights on the three tallest peaks and some faint illumination, like blurry little white squares, at the base. Closer up the blurry little squares became huge doorways—the entrances to the tunnels into the hollow mountain.

At three o'clock in the morning we parked less than twenty yards from the concentric chain link fences surrounding Manzano Mountain. My driver and guide blinked his headlights in a mock code that glittered on the razor wire topping the fences and bounced off the hillside. Thrill-seeking high school kids used to go back into Coyote Canyon at night and turn flashlights toward the tunnel entrances, waiting for the guards to shine huge floodlights back at them. We expected a response at least that dramatic to prove the vigilance of the Manzano Base sentries and the value of the (officially unconfirmed) stuff they guard.

Ten minutes passed. I got out of the car and wandered through the parking lot of the Central Training Academy, where the Wackenhut Company trains the security forces for the Department of Energy. It was a typical high desert midsummer night, cooler by twenty degrees than the midday temperatures that can fry eggs and addle brains. Another five minutes elapsed before headlights loomed in the distance. I squatted down. It's a good thing we weren't spies. The guy in the Air Force security jeep drove right past us, either oblivious to or unconcerned by the sight of me frozen on my knees in the parking lot.

•   •   •

Sandia National Laboratories is protected by its own tall security fence half a mile inside the Wyoming Gate to Kirtland Base. The few outlying technical areas are standard stops on the standard guided tour, which I took with seven minority science students from a local junior high school. They were supervised by a tired-looking young English teacher who sponsored the science club because no one else would. Inside the visitors' center, a putty-colored geodesic dome the size of a small house, our laboratory guide showed a short orientation film, "Working Today for a Secure Tomorrow." It opened with a few seconds of light elevator music and shots of aboveground nuclear weapons tests, bombs floating gently to earth under shining parachutes, and other ominous images.

As the narrator explained the serious work done at Sandia the ninth graders flirted and giggled. "Minority" in New Mexico generally means Hispanic. Just under 40 percent of the state's population is Hispanic. Native Americans make up about 8 percent, and blacks only 2 percent. Non-Hispanic white people ("Anglos," whether WASP or not) are thus a very slim majority. The four girls and three boys were all Hispanics, dark-haired, dark-eyed, some with the slightly flattened noses and epicanthic eye folds hinting at Indian relatives somewhere in the family tree. When the lights came up our escort herded us onto a minibus and we headed for the Photovoltaic Facility, a bank of solar cells that did nothing to engage the interest of the kids.

Sandia is big on alternative energy research, the guide claimed, although there had been more money and hope for that field during the Carter administration. Later we would see more of these public displays—another array of photovoltaic cells that generates about thirty kilowatts of electricity (enough for thirty average homes); a "Power Tower" that focuses the heat from 222 computer-controlled mirrors onto a 250-foot-high column of pipes filled with molten salt; a huge wind turbine not working that day. The middle-aged man droned on about alternative energy sources while the kids stared vacantly at the glass-covered panels and elbowed each other with private jokes. Suddenly, bang! Our guide jumped. One of the solar researchers made a phone call from his office in a nearby trailer and reported back that the explosion was a test of some part of the Midgetman system, a proposed replacement for the MX missile. The teenagers started to look hopeful.

Our bus got held up by Air Force traffic controllers. Something was on its way to Manzano Mountain. The kids had resigned themselves to further energy displays. Things became interesting, though, after the convoy of olive-drab trucks had passed and we pulled up to the older of Sandia's two particle beam fusion accelerators (PBFAs). These accelerators generate beams of ions. In theory, although not yet in practice, when focused on a small area they can cause momentary fusion (at a cost of about $40,000 a shot). It was the last working day before Halloween and the PBFA staff members were in costume. Devils and witches were returning from lunch. A stocky redheaded man was dressed in a green leotard and tights with vines and leaves twined around his body. The old Earth spirit did a sensuous little dance in the parking lot before disappearing into a computer control room.

Eventually we were driven into Sandia's Tech Area One, inside the security fence. After four hours, the tour was finally headed for where the work is done today for a secure tomorrow. The smiling woman in the security booth waved at the kids and counted out a handful of plastic red-and-white candy-striped visitors' badges. Sandia employees call the outlying research facilities where we had spent so much time the "leper colonies." New scientists at the lab are stuck working on unclassified projects outside the fence while they wait for their security clearances to be issued. Those who work at the Power Tower are so segregated from the main research area that they go for their lunch at the cafeteria in the Inhalation Toxicology Lab. They call it the "Puppy Palace" and don't want to know the dogs' fate.

The Sandia field trip took five-and-a-half hours. There were ten stops. Nuclear weapons were mentioned three times. The orientation film had described Sandia Laboratories' national defense mission: to develop and test every nuclear weapon in the arsenal; to train the military to handle and use these weapons; to develop methods for safeguarding nuclear materials, including bombs; and to monitor the condition of the weapons in stockpile. The subject was then dropped until the end of the tour. At a wind-tunnel demonstration an earnest engineer with a salt-and-pepper crewcut introduced the future scientists to aerodynamics with little models of planes and missiles, reminding them that nuclear weapons are the prime mission of the laboratories. On the way out we visited a seismic monitoring station where scientists interpret the pattern of earth tremors recorded by seismographs on long strips of paper. The tracings reveal both earthquakes and underground nuclear explosions, a geologist explained, without showing us the difference.

What the students will probably remember about Sandia National Laboratories is the glassblower. The Glass Formulation and Fabrication Center is a small, clean, airy shop with a half-dozen or so workbenches lined up against the walls. The glassblowers at Sandia design and produce containers and parts when researchers need a particular shape or quality in a glass object that is not available through commercial suppliers. The slender Hispanic man who demonstrated his art was clearly pleased by the chance to perform. He blew glass bubbles, compressed glass springs, and pounded glass beakers, all the while explaining the marvelous versatility of the frozen liquid. He had decorated his workstation with family photos, picture postcards, and a sign reading, "Are You Prepared to Meet God?" The kids were delighted when he asked one of the girls to hold one end of a pinch of glass in a pair of pliers and then backed himself halfway across the room, stretching a delicate trail between them.

• • •

Far back in the Manzano mountain range a dozen old tanks lie rusting and rotting, their insides gutted. Sometimes the military will drag obsolete heavy equipment across a firing range so soldiers can practice with live rounds, but these tanks, their hulls unmarred by bullet holes, must have served some other purpose. Number 13E879 had a piece of paper in it, Copy Number Three of the Equipment Control Record. In early 1983, the tank had been shipped from Anniston Army Depot in Alabama to the Sandia Corporation.

The Sandia Corporation? Sandia National Laboratories has had several names. The older men there habitually call it "the Company." It used to be identified on blueprints as "Project 5-Y," an offshoot of Los Alamos Laboratory, itself code named "Project Y" during World War II. Wise guys inside the fence call it "Sandia Bomb and Novelty." The aliases do not hide the third largest employer in Albuquerque (only Kirtland Base itself and the public school system have more people on their payrolls), but they do reflect the laboratories' indecision about public relations. Since the end of the war Los Alamos has been synonymous with the atomic bomb. Outside of the state and the military-industrial complex, Sandia National Laboratories is known, if at all, as an engineering outfit of some unspecified sort.

The scientific and engineering researchers at the two New Mexico weapons laboratories are divided into two groups, or "series": the professional scientists and engineers, called "MTS" (members of the technical staff) at Sandia and simply "staff" at Los Alamos; and their assistants and inferiors, the "techs." Each lab also has two types of managers. Professional and career administrators work in nonscientific departments such as Personnel and Public Affairs, while those who run the research and engineering projects maintain their staff member status.

The support personnel at the two labs (accountants, clerks, secretaries, maintenance workers, etc.) have their own series designations. Some are unionized and some work for private firms under contract to the labs. But the two laboratories' "real" work, their weapons work, is done by scientists, mathematicians, and engineers with the aid of various types of technicians. As a federal contractor, Sandia's management is motivated to encourage girls and minorities to learn science and engineering. At Sandia, less than 10 percent of the members of the technical staff are black, Hispanic, Native American, Asian, or female. Minority and female representation goes up as job status goes down.

The professional staff members at Sandia and Los Alamos enjoy a status comparable to university faculty members or staff physicians in a hospital. Only those who were frightened by a series of layoffs in the early 1970s express any doubts about the security of their jobs. But there are important differences between the two labs. Most of the 2,600 professionals on the technical staff at Sandia are engineers—electrical, mechanical, nuclear, chemical, aeronautical, civil, materials science, and so forth. In contrast, most Los Alamos staff members are in the so-called basic sciences, primarily physics, but also mathematics, chemistry, and computer science. More than half of the Los Alamos staff members have Ph.D.'s. Only about 40 percent of their counterparts at Sandia are as well endowed with advanced degrees.

What does it mean that Los Alamos does science and Sandia engineering, apart from the fact that more researchers at Sandia are engineers by training? For one thing, Sandia Laboratories is simply more regimented—some say uptight. "They used to have TV cameras on the parking lot to see how many people came in after eight o'clock," complained one former technician who decided she'd rather quit and work as a private consultant than sweat over a decision about whether she had time to go back to use the bathroom after a second cup of coffee at breakfast. Those cameras helped lab officials compile their Parking Lot Index, a part of their internal security program. They monitor who gets into work late, but are more interested in learning of people who routinely stay past normal working hours.

All the buildings at Sandia are numbered. So are the scientists' and engineers' working groups. Each is designated by a detailed code that, once mastered, reveals in a general sort of way what type of work someone does. My copy of the Sandia Labs phone book (which contains multiple official warnings against unauthorized possession and was graciously provided by an engineer grown weary of so many rules) happens to be from 1983, the year Armored Assault Vehicle #13E879 was sent to its mysterious fate at the lab.

The tank was addressed to a man with the organization code number 7535 after his name. That means he worked in Organization 7000, one of seven organizations at the time, each under its own vice-president. 7000 was Technical Support and, typical of other organizations at Sandia, had five directorates under its control. Directorate 500 controlled Development Testing. In Development Testing, the tank recipient was in Department 30, Division 5—thus 7535.

Departments may include as many as seventy members organized into divisions. Divisions are small working units, typically with five to fifteen people in each. The Development Testing Department contains divisions devoted to environmental and field simulation tests of the weapons and devices designed by all three weapons laboratories. The parts that make the guts of a bomb usable are designed in an organization (#2000) innocuously titled Component Development.

In an institution where everything and everyone must be accounted for at all times this would seem to make it simple enough. But the organization charts for both Sandia and Los Alamos are out of date almost before they are printed and distributed. The first Sandia organization chart I saw, for example, had replaced one issued only ten weeks earlier. It too was quickly superceded by another. Both laboratories undergo seemingly constant restructuring, purportedly in response to shifts in their budgets and research programs, and because of their managers' sincere desire to improve communications and productivity. As a side effect, it becomes very hard to keep track of who is doing what, with whom, and under what administrator. Employees' work histories are bizarrely convoluted. The reorganizations confuse laboratory workers, who spoke with thinly disguised scorn of their managers' talent for "taking perfectly good working groups and breaking them up." One technician speculated that personnel are frequently reshuffled simply to confuse Soviet spies.

● ● ●

Tech Area One is big. The employee parking lot (cameras no longer in evidence) is outside the fence, so during business hours the compound is filled with frenetic foot traffic. Some Sandians navigate the maze of buildings in small electric cars, the remnants of a project that was hot at the lab in the 1970s. Back then cuts in weapons research funding led to what one administrator characterized as a frantic effort to diversify into peaceful energy research—on wind and solar power, nuclear power plants, and electric cars. When money for weapons increased, the electric cars became dinosaurs. They still run, but the responsibility for maintaining them now rests with a scientist who works mainly on neutron devices, an important component of nuclear weapons.

The fate of the electric cars and other research on alternative energy sources illustrates a peculiar feature of the national laboratories. Both Sandia and Los Alamos have the potential for nearly infinite flexibility. When alternative energy was popular, they did alternative energy research. When President Reagan called for Star Wars, they did Star Wars. When relations between the United States and the Soviets started to thaw in the late 1980s, the laboratories intensified their research on technologies for verifying treaties. Although laboratory researchers worry about their projects (and sometimes their jobs) whenever there is some major shift in U.S. national security policy, the laboratories seem to be skilled at accommodating and taking advantage of every shift in political climate.

From the third-floor fire escape of a chemistry laboratory, Sandia's Tech Area One looks like a mismatched jumble of army-drab and off-white "temporary" buildings (long since accepted as permanent) planted in rows on a checkerboard of gravel and cement. Concern for function has historically outstripped any appreciation of form at Sandia, although recent years have seen a few cosmetic improvements. The people responsible for new weapons development work in a renovated building. There all traces of prison architecture have been erased by a skillful interior decorator. Plants fill an atrium under a central skylight. The offices open directly onto the garden court or the balcony circling the second floor. Most of the solar engineers formerly housed in easily accessible trailers and temporary buildings outside the fence have also been moved into nice new quarters in Tech Area One. The bulk of the renovations has been done since 1980, when Reagan took office and the outlook for the labs became brighter for those not committed to the development of electric cars.

Despite attempts at beautification, the single most common "decoration" at Sandia is a large metric conversion chart. It graphically illustrates in primary colors how many inches to the meter, the equivalence of metric and nonmetric volumes, and so on. An older fellow in an electronics laboratory I visited smiled when I asked him about the chart, which seemed to have a prominent place in every office in that building.

"The lab printed about a million of those charts when they were thinking of converting to the metric system," he said dryly. "And every single one contains a technical error in definition."

He and the other staff member in the room stopped laughing when I asked what the error was. Hmmm. Neither could find it. It didn't really matter, though, since Sandia's experiment in metric conversion was short-lived and perhaps insincere—its first metric project was called the "Eight Inch Shell."

Those lucky enough to work in redecorated buildings can conjure the gray and green metallic feel of the place just by walking outside. A young scientist working on a nuclear waste disposal project, one of the unlucky ones stuck in a squared-off version of a Quonset hut, was envious of the upholstered chairs in a study lounge at the University of New Mexico where he'd come to tell me about his work. Why couldn't Sandia workers have matching chairs and carpeting in their offices? he wondered.

"There are days I walk into that gate and say, 'I've just got to get *out* of this place or I'm gonna go nuts.' The problem can be summed up in one word—ambience. It's just crazy to go to work in a place with barbed wire, guards, electrified fences, machine-gun nests. I find it really oppressive. And the physical plant reinforces that completely. It's grim. There's not a blade of grass, not a living cell, anywhere in the place."

He hesitated before continuing. "I think there's also a psychological ambience. Although I don't confront weapons stuff on a day-to-day basis, every so often a forklift will go by with a missile or something on it and you'll go, 'Oh my God, what am I doing here?' "

•   •   •

Even the bright lights of Albuquerque were invisible from the back side of Manzano Mountain. The gravel road around the mountain was cut off by a gate high enough to drive under, which we did, just to be sure. The gate lies beyond a picnic area beside a shallow gully. The single concrete picnic table is seldom if ever used. Hearsay has it the tiny spring nearby is contaminated with arsenic.

At first Manzano Base seemed abandoned. Then we saw the lights of a vehicle moving inside the perimeter fence. Gradually a pattern emerged. Every five minutes headlights appeared, seemingly from nowhere, wound their way in S-shaped curves up and down and across the face of the mountain, then disappeared again, just as another vehicle started to snake its way along the same path from the other side. On every other pass around the mountain the headlights met and crossed paths. Even with the wind in their favor, in the stillness before the dawn the voices of the guards at the tunnel entrances carried down the slope and across the gully. They were almost loud enough for us to make out the words.

An untrue local rumor states that no pictures can be taken of Manzano Mountain, that its caretakers have it rigged so unauthorized photos develop black. Supposedly the tunnels twist and snarl for miles inside the mountain. Supposedly the vaults in there are filled with cars, television sets, and other household appliances entombed as part of a long-term experiment on how radiation affects ordinary machines. Supposedly it is the largest storage site for nuclear weapons in the world. I heard this from several Sandia employees in a position to know for sure, all of them aware that the official silence about Manzano Mountain doesn't fool or reassure people in Albuquerque one bit.

One of Sandia Laboratories' permanent responsibilities is nuclear weapons reliability testing. A reliable weapon (what they call a "good one") goes off only when it is supposed to go off. Too soon, and the bomb is "premature." Too late (or never), and it is a "dud." For a bomb to be reliable, its parts must be reliable, and they must operate as their designers intended even after they have sat for years in stockpile, been banged around by GIs, gotten dusty or damp, or suffered any other conceivable indignity.

A nuclear weapon is not like a Volkswagen, which can sit in a garage forever and still run fine once you put a little air in the tires. Conventional explosives, radioactive materials, and highly sensitive electronic circuits are in close proximity inside a nuclear weapon. Quality assurance is more difficult when you're talking about such complicated devices, Owen explained.

Owen was a jovial mathematician with an infectious grin. Now in his mid-fifties, he had worked at Sandia for fifteen years, but his career as a government weapons engineer began when he served as an electronics technician in the Korean War. The closest he had come to battle was courtesy of an irate fisherman in Pusan Harbor who shot at his ship—"Well, it was a .22," Owen admitted with a laugh. Ending up with a Ph.D. in mathematics surprised him, because in high school, to avoid being too much of a conformist, Owen said he had very carefully maintained a D-plus/C-minus grade-point average. But after his hitch in the service one electronics job led to another, and eventually he was hired at Sandia to keep nuclear weapons "healthy." The weapons lab put him through graduate school while he worked.

America's nuclear weapons are stockpiled for twenty to twenty-five years. "After that long, metal grows—it grows whiskers. And the whiskers can grow through wire insulation," Owen continued, "and can close a circuit." Prematurely closed circuits upset the carefully designed equipment. To prevent this, Sandia researchers design circuits that are "hardened" against radiation. They periodically take a sample of the missiles in the arsenal to check for whiskers and other threats to reliability. Owen figured out how to make those checks more efficiently. "I was very proud when I helped cut down the sampling size in quality assurance because it helped to save money," Owen said. "And my feeling is if we keep the old ones alive we won't have to build new ones."

It is useful to be able to predict what will happen when nuclear weapons grow old, so some divisions at Sandia do accelerated aging studies. Weapons researchers put the relevant parts of nuclear weapons into contrived environments that simulate the stresses caused by the passage of time. In essence, they put a bomb into a time machine that's headed for the end of the bomb's lifetime.

Then they compare their predictions with the facts. That reportedly and probably is the major purpose of Manzano Mountain—to house a supply of bombs that can sit and grow old naturally or be recruited as time travelers.

• • •

Sandia National Laboratories is known as a "good citizen" in Albuquerque. People who work at the lab are a political force in the community. A lot of them vote. Some hold public office. Their United Way Fund Drive is always a big success, and many of the people who told me about their work at the lab also talked about their volunteer work, for the Boy Scouts, the church, the Big Brother program. The lab sponsors some of the science programs on the local public television station. "Albuquerque was a small town when Sandia was started in the late forties, and there was a mutuality of interest," one man reminded me. "Sandia people have become integrated into the community. Albuquerque and Sandia grew up together."

They test the air-raid sirens in Albuquerque at noon on the first Tuesday of each month. When the siren near a big Ford dealership short-circuited in a rainstorm late one afternoon and wailed for forty minutes, hardly anyone changed their schedule, although one of my neighbors sent her teenage sons down into the basement. Every spring Air National Guard A-7 fighter planes in formation swoop low through endless practice runs over the University of New Mexico campus. Students sometimes feign fear or shake their fists at the noise, but after a while they get used to it. Each morning T. J. Trout, a popular loudmouthed deejay on a local station, "blows somebody up" on the radio. You can nominate your boss or some local luminary to serve as the target of his atom bomb.

The Federal Emergency Management Agency identifies six locations in New Mexico as being at "high risk" for nuclear attack if deterrence should fail. Albuquerque tops the list. A few years ago the then-director of the city's civil defense office outlined the emergency plan at a public forum organized by a group of local citizens.

Standing before a huge map of the city, he placed the tip of his pointer on the cross formed by the intersection of Interstates 40 and 25, known locally as the "Big I". "We figure the Soviets would drop two bombs right here," he said, tracing the circumference of two yellow circles on the map and describing what would happen next. In a nuclear attack, those within a two-mile radius of ground zero could be considered lost. Survivors outside the yellow circles would be instructed by radio and television to circumvent the blast zone on local roads and drive north to Santa Fe. There they would take shelter in the basement of the open-air opera house and be fed by local restaurants.

"Why the Big I?" someone called from the audience.

"Because these are the two major highways, and by bombing the intersection they could cut off the major truck routes through this part of the Southwest," he explained patiently.

"But while they were at it, wouldn't they also bomb the Air Force base and Manzano Mountain?"

"Well, we don't think so," he said promptly.

"Why not?"

"Well, we just don't think they would," he said. The audience was silent.

•   •   •

We sat watching the jeeps and listening to the sentries on the back side of Manzano Mountain for nearly half an hour before a DOE security guard came by to unlock the gate. The guard admitted that the area was not restricted but advised us to leave anyway, slapping the side of the car hard to be sure we took the hint. Just then a jeep roared up and through the gate, so fast it raised a high column of dust that obscured a section of the stars. As we drove away the headlights continued their unending circle dance, like perverse fireflies unable to resist the darkness of the mountain.

3

THE MAGIC HILL

.

.

.

Los Alamos means "the aspens." Aspens are poplars, slender, fast-growing trees in the willow family. Aspen leaves are thin and light, attached to their branches by the slimmest of stalks. Like living hearts, they are never still. The slightest breath of air makes them shiver and shimmer.

The town of Los Alamos itself has a certain shimmering, hallucinogenic quality. It sits 1,300 feet above the Rio Grande Valley—and 7,400 feet above sea level—on a fan-shaped mesa in the Jemez Mountains. The high alpine plateau is cut by deep canyons, its topmost surface divided like islands in a stream or fingers on an outstretched hand. At night the lights of Los Alamos shine so far above the normal horizon the city appears to be moored among the stars.

From Albuquerque to Santa Fe to Los Alamos travelers encounter the terse warning "Speed Monitored by Aircraft," universally recognized as untrue and irrelevant. Almost everyone speeds through the vast expanses of the West. But the last leg of old State

Road 502 to Los Alamos ascends the red and yellow mesa with turns so steep and clearances so narrow it forces all but the most foolhardy to slow down. Small cars and fainthearted drivers are equally taxed these last few miles, not least because the oncoming downhill traffic moves at near light-speed. The most reckless drivers are supposedly the teenagers headed down to the poor valley town of Española where they don't always check ID cards before they sell liquor.

By the time you recover from the final hairpin turn past a sheer vertical cliff, SR 502 has become Trinity Drive, the posted speed limit is thirty-five, and you officially enter the birthplace of the bomb. Los Alamos is a high, glimmering mirage when glimpsed from below at night. Thinking of the Manhattan Project scientists as modern counterparts of Prometheus, one author called it the "City of Fire."[1]

The main street in Los Alamos is named Trinity Drive, after the spot where the first atom bomb was exploded. Oppenheimer claimed in a letter to General Leslie Groves, the army head of the Manhattan Project, that two poems by John Donne were in his mind when he suggested "Trinity" as the code name for the bomb test site.[2] Oppenheimer said he was not sure what had inspired the choice. Surely people can be forgiven for assuming the name was intended to conjure an image of God. But the eighteen thousand or so residents of Los Alamos County—Los Alamos itself and its tiny "suburb" of White Rock—simply call their home "the Hill."

The nickname dates to the community's founding. The secret town was really an army camp with room for the wives and kids of the civilian scientists who showed up ready to help turn a four-year-old discovery—fission—into a bomb. The entire county—New Mexico's smallest—was created by the army, cut out of two surrounding counties to ensure the security of Project Y. Calling Los Alamos the Hill no doubt helped keep the veil of secrecy in-

[1] James W. Kunetka, *City of Fire* (Albuquerque: University of New Mexico Press, 1979).

[2] Richard Rhodes, *The Making of the Atomic Bomb* (NY: Simon & Schuster, 1986), pp. 571–2.

tact. They still design bombs in Los Alamos, but now the place is known to schoolchildren throughout the world. With the Parajito Plateau far too high and astonishingly convoluted to be considered a hill, the old nickname seems a bit of false modesty.

On the other hand, "the Hill" has a certain homey ring to it. And up on the mesa, gawkers are disappointed. Wondrous-sounding Trinity Drive is an ordinary asphalt strip. It runs by a shopping center that could be in Anytown, U.S.A., with a grocery store, a drugstore, and a good bookstore. It takes you past a couple of Howard Johnson–style motels. To the left is a McDonald's and a gas station. To the right, round little Ashley Pond, once the swimming hole for the Los Alamos Boys' School, sits primly, ringed with cement, in the middle of a carefully tended emerald lawn. On this part of the mesa the Army Corps of Engineers erected ugly but functional temporary structures for the wartime Los Alamos Scientific Laboratory. Scientists were lured to the new laboratory "on the shores of a small lake," as the pond was referred to in a recruiting brochure. After the war, the building began in earnest, producing permanent facilities for the laboratory and a boom in upper-middle-class housing. Now almost every building in town was built after 1945.

The Southwest has a distinctive architecture. The "Santa Fe style" is a natural consequence of construction with soft adobe brick. Like their old Indian and Spanish models, modern concrete block buildings end up with gently rounded corners, interior archways, and walls stuccoed in a clay-and-sand palette—pinkish tans and creams, dusty oranges and browns. It is a peasant architecture, always romantically on the verge of cracking and becoming mud. You see it everywhere in New Mexico.

Except in Los Alamos. Los Alamos architecture is mostly clean and crisp. Buildings here have sharply defined edges. Looking like a Colorado mountain ski resort, apartment complexes along Trinity Drive have natural wood siding. Trim one- and two-story single-family homes sided with brick, wood, or aluminum line streets that curve along the edges of the irregular spits of land between the canyons. With so little of New Mexico's distinctive architecture, Los Alamos looks even more like it belongs nowhere in particular.

Traffic stalls on Trinity Drive twice each weekday, for fifteen minutes in the morning and again when the laboratory workers head home. Convoys of joggers hug the roadsides at lunchtime. But there are no gleaming white towers, no inspiring statues of the great physicists, no American version of the Arc de Triomphe. There is no visible sign of the bomb at all.

Hardly a bit of litter intrudes on Trinity Drive, much less a beggar. When Trinity Drive narrows and heads into a residential neighborhood, laboratory traffic takes a left onto a graceful steel span over the Los Alamos Canyon. At regular intervals small signs on the four-lane bridge warn of your entry onto government property. Only pedestrians or passengers with binoculars have any hope of reading the details. Government property isn't news in Los Alamos, anyway. A mere 5 percent of the land in the county is privately owned. The federal government (mostly the Department of Energy and the National Forest) owns about 90 percent. The postwar Los Alamos National Laboratory has spread itself over forty-three square miles on the top of the mesa.[3]

The Winnebagos and camper vans with out-of-state licenses continue past lab property to Bandelier National Monument and the small shelters carved by Indians in soft white stone cliffs 750 years ago. Signs of ancient inhabitants litter the top of the plateau, including a small remnant of a Tewa Indian ruin at the edge of a shopping center parking lot. Today the Indians live in the valley, on land deeded by treaty to pueblos with Spanish names like Santa Clara and San Ildefonso. The few who work at the lab are commuters.

Getting a look at Sandia National Laboratories requires plenty of advance planning, a good reason to visit, and a guide to take responsibility for you. In Los Alamos, you're on your own. Visitors and employees compare the Los Alamos Laboratory complex to a huge college campus without students. Buildings and laboratories are scattered over dozens of numbered Technical Areas on the

---

[3] Sources are inconsistent. *NM Statistical Abstract* claims 93 percent of the land is federally owned; the Los Alamos Chamber of Commerce claims 88 percent.

southern and western edges of the mesas. People suspecting the worst of Los Alamos immediately notice that most of these areas are surrounded by barbed-wire and chain link fencing. So much fencing imparts a military quality to Los Alamos. It is rumored the trees droop. But the greenery seems perky enough, and the gates to most technical areas stand open, no sentries in view.

No one cares if you ride around and indulge your curiosity. Visitors are free to ogle the lab architecture (mostly 1960s prison style, with no particular coherence or grace), to sit under the pines near the administration building, to squint past a fence at the bunkers where explosives are stored. Basketball hoops mounted in many of the Tech Area parking lots inspire lab employees to get up a quick game during the lunch hour. No gleaming towers or awe-inspiring statues here, either.

Ask people who live there about Los Alamos and they grope for explanations. The place has peculiar characteristics, quickly revealed by a brief consideration of some numbers. For example: In 1980, half the residents of Los Alamos had completed more than fifteen-and-a-half years of schooling, three years more than the national average. Educated people in Los Alamos studied the *hard* subjects. About four out of ten workers in the county are scientists, engineers, or technical specialists. Sixty-nine percent of the Los Alamos High graduates go to college.

Almost everyone in Los Alamos County is white and affluent, with household incomes averaging about twice those in the rest of the state. More than half of the people in the county work for the national laboratory or industries associated with the lab. Altogether, the government sector, including the Los Alamos consolidated city/county government, provides three-quarters of the jobs on the Hill. The one company in this town is Government.

Los Alamos is a recognizable stereotype, the kind of town where most Americans probably suspect everyone *else* grew up. Here the bourgeoisie reveal their charm. Having been associated for so long with racial prejudice, stereotyping has gotten a bad name. Accurate stereotypes are useful, though. They help you guess what to expect in new situations. Los Alamos is stuck with its reputation as the home of great accomplishments with fearful consequences. It is the town that ushered in the present. But Los Alamos does not embrace the honor. The reality here is a mundane version of the 1950s suburban dream. There are two cars in the driveway of the comfy house with the neatly trimmed lawn. Dad is a professional. Mom may work; she helps out at the church. Sis is on the swim team. Brother drives a neat car and gets into scrapes with his pals. When you ask Brother about his hometown, he thinks of his high school days.

"There was no escaping it," Gene said, leaning back in the wooden booth in a noisy Albuquerque bar. He frowned at his pale hands, the fingernails bitten to the quick.

He had sharp blue eyes and an even sharper profile. "I could leave Los Alamos and go to Massachusetts and people would say, 'Where are you from?' You say 'Los Alamos' and there's all these questions about it. It feels—it's like being proud and ashamed at the same time."

Gene graduated from Los Alamos High and worked as a technician in the national laboratory. Fed up with his bosses at thirty, he decided to finish his bachelor's degree. With a business major he could be his own boss.

Like high school kids everywhere, Los Alamos students have cliques and crowds. When Gene was in school, most kids were either socializers—heavy drinkers who liked to party—or nature lovers who hiked for fun. Gene liked parties and camping, but he *loved* science.

"I was a Conehead," Gene confessed, grinning as he lit a cigarette. The Coneheads, a television family of pointy-headed extraterrestrials, drank six-packs of beer and had weird misadventures on "Saturday Night Live" in the 1970s. They were scientifically sophisticated, completely humorless, perfectly literal, and unfailingly logical. The Los Alamos Coneheads inherited their fathers' interest in the technical working of things. Some also drank beer. Twirling the suds in the bottom of his mug, Gene admitted an adulthood preference for marijuana.

As they do everywhere, the kids in Los Alamos pull pranks. The tradition began long ago, with a generation driven to rebellion by a ten o'clock curfew, no movie theater, leagues hogging the bowling alley on weekends. Boredom begat diverse expressions of resentment. The grown boys are happy to list typical Los Alamos high jinks. Like, drive down the street spraying bystanders with stolen fire extinguishers. Ring and run. Watch drivers' reactions to paper-stuffed dummies in the street. Egg things.

The pranks become truly inspired during Senior Week, as each year's senior class competes with the legacy of the previous year's tricksters. The Coneheads mastermind the most technically sophisticated tricks. One bunch maneuvered a huge metal flagpole into a small room. The school administrators, failing to think "skylight," scratched their heads and resorted to a blowtorch, cutting the flagpole into chunks they could get out of the building.

Despite the barroom din Gene lowered his voice. He leaned across the table to explain how *he* had linked all the clocks in the high school to the phone system. Each time a principal's phone was put on hold a light on the phone would blink. Each flash advanced clocks throughout the school one minute. "It drove them nuts! They couldn't figure it out." Gene's grin grew wider with the memory. Time at Los Alamos High became relative, a function of communication. For days all the classroom and hallway clocks would sporadically cycle five, ten, twenty minutes ahead. Then resume their normal pace. Then speed up again.

"This town is the biggest change maker around," one man said. Somewhere the profundity should be visible. But on top of the mesa the history is practically invisible. In a small building near Fuller Lodge, one of the few buildings remaining from the old Los Alamos Ranch School days, the Historical Society remembers the past. The Society has it all preserved in a dim room—the texts, the photographs, the artifacts from the beginning of the nuclear age. Outside the door, however, the spell is broken. It hardly seems possible that anything remarkable ever originated here. Los Alamos is not a living museum.

"I think the people struggle with the same issues here as in any similar small town," said Dale Arnink, minister to the 150-odd members of the Los Alamos Unitarian Universalist Church. The standard joke has it that a Unitarian is an atheist who can't break the Sunday morning habit. Churchgoing is a big pastime in Los Alamos. The Chamber of Commerce hands newcomers a mimeographed guide to the county's thirty-two congregations. The Unitarians are reputedly more liberal than their neighbors. "Liberal" is a relative concept, however. The town is registered solid Democrat. But in major elections, the vote is invariably two-to-one Republican.

"Because we're so isolated, people *feel* that makes the community different," Arnink continued. An attractive and easygoing man in his late forties, Arnink's beat-up recliner was right in the middle of his tiny study, the location apparently permanent by necessity. Except for a narrow path from door to chair, the floor was covered with papers and books. Pieces of ivy rooted in water glasses balanced atop leaning towers of paper. From his chair Arnink could reach everything without risking an avalanche.

Arnink had once taught high school in a California university community with some high-tech industries. They had the same kinds of problems there, he recalled, problems of affluence and middle-class living.

"I have a friend who's a Jungian analyst and she's convinced that the shadow side of this town is really terrible, that a place that is working on something that can destroy the world must have a very dark shadow side," he admitted. "There's some logic to that. But I just don't see the evidence for it." When the kids take drugs, it is not because they are in Los Alamos, home of the bomb. It is because that's what teens do in affluent smallish towns. Their parents get angry at the lab, but not because the lab designs nuclear weapons. They are frustrated, Arnink thinks, that their opportunities depend so completely on one institution. The one-company town is especially cruel to the well-educated women whose master's degrees in English literature qualify them for nothing more challenging than a secretarial job at the lab. But there is no shadow.

This opinion is popular in Los Alamos. Tourists expecting something remarkable from the town hear it over and over again: "It's just an ordinary small town." "It has the problems of any small town dominated by a single industry." When outsiders offer an interpretation challenging its Anytown, U.S.A. quality, people in

Los Alamos react with anger and resentment. They ask if you've read a ten-year-old *Time* magazine article that found modern-day Los Alamos unusually conformist, repressed, and unhappy. "How could *Time* have done such a hatchet job?" they wonder. They found Vivian Gornick's more recent story, "Town Without Pity," infuriating.[4]

And why not? The younger generations of scientists have no compelling reason to adopt the historical burden left on the mesa by their Promethean predecessors. It gave them their jobs, but they didn't start the nuclear age. The war years in Los Alamos were remarkable, full of purpose, energy, camaraderie, and creativity born of necessity, but the war is long over. The old excitement has been replaced by comfortable routine.

One child of the war years remembered a scientist on Trinity "wearing one sneaker and one black dress shoe, riding a bike in the rain reading a book." "From the forties to now they haven't changed a bit," he said. Feeling unappreciated and put down by snobbish Ph.D.'s, laboratory technicians quickly accuse their higher-status professional neighbors of being "blue-sky guys," with their heads in the clouds and not enough sense to come in out of the rain. But no egghead eccentrics are visible on Trinity Drive nowadays. In their stead are normal-looking fellows in sensible cars who obediently stop for panting joggers in warm-up suits. The strange idiosyncratic geniuses are gone, or disguised as conformists.

The cadre of dedicated and energetic atomic scientists committed to a single monumental task of profound import is gone, too, dead or in retirement. Newer residents show little interest in the history of Los Alamos. Public talks by the older generation about the war years draw dwindling crowds. That old reality is now an origin myth. It no longer has a tangible form on the mesa.

After convincing you that Los Alamos is a normal sort of place, people feel free to complicate the story. The Unitarian minister strenuously defended his contention that the town is typical. But

[4] Joseph Kane, "Los Alamos: A City Upon a Hill," *Time*, Dec. 10, 1979; Vivian Gornick, "Town Without Pity," *Mother Jones*, Aug./Sept. 1985, p. 14.

after an hour Arnink began to consider the subjective consequences of those objectively peculiar facts about Los Alamos. By
genetic predisposition, early childhood experience, or subsequent
training, he said, most people in Los Alamos believe the scientific
method is the best way to approach problems. Shifting in his chair,
he summarized their simple faith: "All problems have a technological fix." Then he laughed.

The belief that all problems can be clearly defined, systematically analyzed, and rationally resolved makes his marriage counseling sessions a real chore. In a bemused tone Arnink reconstructed the familiar scene. The unhappy couple comes into his
cluttered study for a talk. They need help. The husband is a scientist at the lab. The wife is angry and frustrated. The scientist
suggests they follow a rational procedure. First they should define
the problem. Then they'll analyze it. Then they'll fix it. Arnink
suddenly looked weary. He has to *explain* empathy to these men.

When atomic bombs were dropped on Hiroshima and Nagasaki in
1945, the scientists who had labored to create them were suddenly
thrust into the public eye as national heroes and fathers of a new
age. The light of the explosions more than adequately illuminated
the purpose of their top-secret work. Oppenheimer, head of the
laboratory hidden on the Hill, reported that witnessing the first
successful atomic test in the desert of southern New Mexico
brought to his mind a quote about a Hindu death god, the "shat-
terer of worlds." Forty years later, he and his colleagues from the
Manhattan Project have become symbols of the ambivalent power
of science. Science seems to involve not just what is good and bad
in human beings, but the extremes of each.

"They don't really have much sense of what makes human societies or politics or communities work or not work," Arnink concluded. But if the scientists and engineers in Los Alamos have a
weak sense of the human heart, it is not from laziness. They just
approach the heart in a peculiar way deliberately cultivated by
scientists. They try to be objective.

Objectivity means taking a neutral, open-minded, and disen-
gaged attitude toward things. However profoundly moved by a

new discovery, scientists must adopt a dispassionate attitude in their descriptions and analysis. This can make scientists seem deficient in moral and aesthetic sensibilities.

You can see it in the articles in the world's two most prominent science journals, *Science* and *Nature*. The elegant but arid writing from the prescribed emotional distance of a million miles betrays no joy in research or love of subject. The astronomer's description of the night sky, or some small portion of it, is not poetry. The rules of science are meant to enforce objectivity. Achieving an objective attitude is difficult. But once you get the knack it becomes a hard habit to break. No wonder scientists often seem to botch tasks that require expressiveness and emotional sensitivity.

Every ten years the consolidated Los Alamos City/County Council commissions a residents' survey. By this means the people on the Hill find out what's on their neighbors' minds. There is great attention to detail in these surveys, as in many things produced in Los Alamos. (For example, the Chamber of Commerce's information packet for new residents includes a chart describing the climate in Los Alamos. It summarizes ten different weather variables, month by month, from 1911 to the present. This is the level of detail people appreciate in Los Alamos.)

According to the latest community survey, people in Los Alamos worry most about the lack of affordable housing and the fact that they live in a one-company town. A dearth of privately owned land on the Parajito Plateau has produced the housing crunch. The immediacy and immensity of nature here disguises the fact that nearly 100 percent of the people on the mesa live in urban areas. Most agree that technicians and other less-affluent people need more low to moderate-income housing. But Los Alamos residents really favor single-family homes, and efforts to provide other types of housing have met with mixed reviews. A few years ago a lower-income housing proposal before the elected Los Alamos City/County Council provoked someone to erect a billboard on State Road 502 reading "Vote Yes on Proposition X—Keep Technicians Out."

Bored with the limits of Merrimac and the other shopping plazas, Los Alamos residents would also like a wider variety of retail stores, more like a city. But only 5 percent would rather live elsewhere.

"I couldn't find anyplace where the living was more to my liking," John Manley told me, explaining why, after years as a university professor, he'd moved back to Los Alamos in the 1950s and become research adviser to the director of the lab. During the war years he had served as Oppenheimer's deputy, his second in command. What he found most to his liking was the free spirit of scientific inquiry at the labs and the resources to pursue it. But he also loved the physical setting, the climate, and the small-town atmosphere. "Probably they say the same thing today," he said, and he was right. Those three things top the objective list of why people love Los Alamos.

The small-town atmosphere lets people feel secure in Los Alamos. One black professional had built a fancy wood entryway for his house. He waited a year and a half before finding exactly the type of lock he wanted. "That was a year and a half our house wasn't locked," he said, in which nothing bad happened. Nothing was stolen. No one bothered his white wife or their four children. He knew that "the rest of the world" wouldn't let them live so comfortably.

The crime rate is low. With less than 4 percent of families living below the poverty line (most of whom are probably elderly retirees), few feel a pressing need to steal. Teenage vandalism is a constant annoyance, but violent crime is rare. Except for innovative and rebellious teenagers, most people in the town are quite staid and civilized. The teenagers' drug use has always worried the adults. They studied the problem with a comprehensive survey.

People in Los Alamos have good manners. They don't try to cut into the cashier's line at the grocery store. They rarely settle disagreements with fistfights. They keep their stereos low and brake for pedestrians.

Many technicians resent the hegemony of upper-middle-class values, the rarefied atmosphere of reason and order. One motorcycle buff, in constant trouble with the "Gestapo police force" that issues tickets if you go five miles over the speed limit, called it "class consciousness." The wives of staff members at the Lab introduce themselves as "Mrs. Doctor So-and-So," he claimed, as a put-down to lowly techs.

"Things are breaking down, though," he reported with satisfaction. "We got a Sonic [drive-through hamburger stand], a McDonald's, and they don't control the radio station anymore. It used to be all long-haired operatic-type music."

But a young physicist doubted the breakdown. The pace of change seemed snail-like to him, diversity nonexistent. "I'm like a zombie here," he complained. Life had become too predictable, imagination and zest overcome by the deadly routine. Rumors kept hope alive. "There's a lot of gossip—if it were true it'd be interesting." But then he sighed. "It's not true."

Adding cruel detail to the stereotype, he described domestic and social life in Los Alamos. During the week, the scientists will calculate how many drinks you can get out of a bottle of liquor. They compare the price of a bottle to the price of a drink at the Los Alamos Inn, and conclude that drinking at home is cheaper. Friday night they go to the grocery store and buy two bottles of gin. Saturday morning they mow their lawns. "They take care of their lawns *a lot*," he said in despair. They go to church on Sunday, even the atheists.

Supposedly religion competes with science for authority on cosmological matters, but you wouldn't know it by the herds of neatly dressed family units walking to church on Sunday mornings. Maybe the scientists are hedging their bets. More likely they are bored. People search desperately for amusement in Los Alamos, and have created distractions in the form of roughly 250 voluntary organizations for everything from astronomy to coupon clipping to folk dancing to chess. Church gets you out of the house. And in a town described by a counselor as "full of workaholics," it provides an intelligent distraction from the problems of office and laboratory.

What kind of sermons do you give to weapons scientists? "This is a sophisticated audience and I can't get away with talking nonsense or making broad generalizations," Arnink said. "When it comes to the scientific mind, I find they're more comfortable with talking about the details and critiquing the details." Arnink tried to set up a panel of parishioners to consider the ethics of war and peace. He got nowhere. Finally, in frustration, he did it himself, publishing the resulting sermons in a pamphlet, "Four on War."

He began the series by praising science and then patiently explaining how facts are often incomplete or unavailable. Without them, behavior is determined by attitudes, which are all too easily set in stone. Objectivity is not always possible or desirable, he explained.

What is scientific objectivity? Why are scientists so fond of the objective stance, even at the cost of their own subjectivity? I think being objective mimics a religious experience. Religion and science both give people a chance to transcend the world in which we take so much for granted. In the experience of spiritual transcendence the self appears to become part of something much larger, more powerful and universal, some great ego-absorbing Other. You get to be part of the river.

Science offers a different route to transcendence. When someone adopts an objective attitude, what exactly is he or she doing? Apart from the soul, if there is one, individuals could arguably be considered little more than organized bundles of biases and conceptions. Objectivity requires putting aside your desires and preconceptions. The attempt to be open-minded, to be completely neutral, is really a profound effort to abandon the idiosyncrasies of the ego. To look at things objectively, you must leave yourself behind. You are no longer at the center of your own perspective. In some real sense, being objective is not human. This may not be the best attitude to take toward intimate relations or war.

• • •

Back in the bar in Albuquerque, Gene was obliging but exhausted by the questions about his hometown. He was uncertain about the facts. Objectively, is it a normal town? Do the statistics really tell the story? Do kids elsewhere in America get impromptu chemistry lessons from Dad over the backyard barbecue? He was confused about the meaning of the facts: Should he be proud or ashamed? Should he brag about it? When he goes out of town should he say, "I'm from Los Alamos, that's where they designed the bomb?"

Should he go out of his way to let people know he is *from* somewhere?

Just as it is hard to enter Los Alamos in a neutral frame of mind, it is hard for people who live there to stay consistent in their feelings about the town. Objectivity supersedes subjectivity, which sneaks back in anyway. Simple stories unravel or become disconcertingly complex. Perfectly clear descriptions begin to shimmer and fall apart.

Consider the two bottles of gin in the grocery bag of the stereotypical Los Alamos scientist. People talk about alcohol a lot in Los Alamos. They say their neighbors drink to excess in the rec room at night. The Hill's head alcoholism counselor denies an unusually high alcoholism rate, believing the highly intelligent residents of Los Alamos more likely to seek help for a drinking problem. Everyone regrets the lack of reliable data on this. In 1974 over 40 percent of the population in Los Alamos labeled alcohol abuse a serious problem. The 1984 citizen survey left the question unasked.

If people in Los Alamos hide their drinking at home, frugality is not the only reason. Counting the bottles of beer on the wall of the bar at the Los Alamos Inn is risky in a town starved for gossip, especially with a security clearance to protect. Loose lips sink ships. Safe scientists drink in silence. Ask about alcoholism in Los Alamos and people declare it a big problem, gossip and speculate, then go on the defensive, their objectivity rattled as if they fear alcoholism reveals a more universal unhealthy impulse, barely hidden and barely controlled.

Then there's suicide. The official suicide rate in Los Alamos County is extremely low. "Everyone says it's high," people say, "but look it up. It's really not." So why does everyone say it's high?

Small-town tragedies always seem more personal. The facts say "Don't worry about suicide in Los Alamos." But people worry anyway, their objectivity fading as they consider how many neighbors were lost to the gun in the mouth. Gene furrowed his brow and ticked off a list of twelve people he knew who had died by their own hand in as many years. Most killed themselves for reasons not generally known to the community. "Self-imposed pressure,"

Gene theorized. "It's a 'science-on-demand' kind of thing—'Today I will be creative'—that they put on themselves." As ideas come slower the pressure increases.

Even though the daily life of the town belies it, Los Alamos may be special because of its history. Maybe it derives a peculiar character from all those scientists enthusiastically and indiscriminately applying the scientific method to everything that holds still for ten seconds. Maybe it is the American dream, dishwashers and clothes dryers in a lovely pastoral setting, comfort and conformity on the Hill. A counselor in town thought hard and suggested that Los Alamos is a metaphor for being an American, "the safest place in the world, the most prosperous, the most beautiful, and yet responsible for the most horrendous things in the world."

It is hard to find the bomb in Los Alamos. It never really shows itself. You can buy books about its creation in the bookstore and see the photos of Oppenheimer, Teller, Manley, and others at the Historical Society. If you want to see what they made you can go to the Bradbury Science Museum, named for a former lab director. There you will find models of the casings for Little Boy and Fat Man. Both bombs have simple boxlike stabilizers as "tails." Little Boy is a fat cigar. It released the force of about thirteen thousand tons of dynamite onto the city of Hiroshima. Rounder Fat Man hit Nagasaki with the equivalent of twenty-one kilotons of TNT.

Identical models of the two bombs can be found in the Atomic Museum on Kirtland Air Force Base in Albuquerque. But the bombs in the Atomic Museum are painted a dull greenish black. They are serious pieces of military equipment. Their Los Alamos counterparts, in an unobtrusive corner of a room devoted to educational displays about atomic energy, are painted shiny white. They are products of science.

Gene grew up during the Cuban Missile Crisis, when kids in Los Alamos conducted a heated debate out of their parents' range of hearing. They used to argue about whether the Soviets would drop a bomb on Los Alamos. Would they be the first to go? Some thought yes. Others said no, the Soviets would want to spare the scientists. It's not clear why the children thought the Soviets would spare their parents. Is Los Alamos objectively too small and peripheral a target? Would they want to keep them alive for their brains? After all, America adopted many Nazi rocket scientists.

"I remember having plans and putting food away in a cave." Gene stared at the dregs of his beer. Understanding the questions of life and death remains a pressing childhood priority on the Hill, Gene insisted, even with the national public school duck-and-cover rituals (two long bells, take cover; one long bell, all clear) long abandoned. "If you asked any kid in Los Alamos, they could tell you where they'd go if there was a nuclear war. They *have* to know what to think about it. They still test the air-raid sirens every week—five o'clock Mondays."

Still, Los Alamos seems heavenly for children and young teenagers. "If you just had a little initiative you could do anything," Gene said. Young people with scientific ambitions go down to Zia Salvage, where Los Alamos Laboratory unloads its obsolete and unwanted equipment. There a few dollars will buy "neat stuff" like computer hardware, energy storage systems, and guidance systems for missiles. What is obsolete equipment from the lab's point of view may be nearly state-of-the-art to everyone else. Once Gene saw they'd jettisoned a laser. On a hunch he combed through the guts of the machine. His reward was a translucent red cylinder half the length and about the diameter of a ball-point pen.

He held it up for me, a bar of pure high-grade ruby the shade of blood in water. "Somebody made a big mistake," he laughed.

4

THE SUN IN
MINIATURE

.

.

.

John Manley fumbled with a strip of movie film so old and brittle
it seemed endangered by even his delicate and competent touch.
The elderly physicist, now in his eighties, had brought them to
Albuquerque from his home in Los Alamos, four small reels of
sixteen millimeter carefully tucked into a plastic bag adorned with
a color picture of its former contents—mixed frozen vegetables.

The audiovisual aides at the University of New Mexico wheeled
a projector into a tiny room so Manley could show the short pieces
of history. There were some home movies taken two generations
ago in the desert of southern New Mexico, before and after the
Trinity Test. And there was a strip of the bombing of Hiroshima.

"Damn!" A piece of the leader had broken off. Manley patiently
rethreaded the projector and focused it on the wall.

Television and film favor particular images of atomic blasts. Everyone has seen them, the black-and-white mushrooms blossom-
ing in perfect symmetry. They are captured by a stable camera on
the ground at a respectable distance from the hypocenter, the

point on the ground directly beneath the center of the blast. The Hiroshima film was taken through the window of the instrument plane accompanying the bomber the *Enola Gay* by Harold Agnew, a Los Alamos colleague of Manley's, who was to later become director of the laboratory. Compared to later films of aboveground atomic tests, the Hiroshima image is poorly photographed, visually cluttered, choppy, confusing. In short, it is less aesthetically appealing, and consequently less likely to be used by the media as a visual cue to the topic of the atom bomb.

Suddenly the image flickered on the wall. Against the backdrop of the countryside beyond Hiroshima a white circle began to vibrate and grow. The city itself was merely implied, lying too directly below the airplane to be captured by the camera held before the glass window. The airplane's window frame cut across the right side of the image, quivering in front of the plane of the landscape. Then, as Agnew shifted position, it disappeared and was replaced by its twin on the left. The screen shook like a B-29. The film is shocking in color. The delta of the river to the Japanese Inland Sea was very green on August 6, 1945.

Most people think of the bomb as one big destructive thing. But to the scientists and engineers at Los Alamos and Sandia, the details matter. Many of them do not think of themselves as working on the bomb, this one big thing; rather, they feel they work on a single small thing, a tiny technically interesting part of a larger project. To see the bomb the way they do, you must see it as a bunch of parts, an orderly process with a predictable outcome.

How do nuclear weapons work? Scientists and engineers from Los Alamos and Sandia respond with blank looks. That seems the sort of query designed to get someone into trouble. The public affairs officer for the Albuquerque Area Office of the Department of Energy was shocked by the question. No one had ever requested that information. He knew of no unclassified description to recommend.[1]

---

[1] For a fascinating account of how difficult it is to get even a basic description of the workings of modern thermonuclear weapons see Howard Morland, *The Secret*

Manley knows how they work. Before the laboratory at Los Alamos was established, he was put in charge of coordinating the experimental work being done on the bomb at the University of Chicago and other universities around the country. He and Oppenheimer together set up the Los Alamos Laboratory, with Manley responsible for "experimental arrangements"—the equipment, the people who used it, and the buildings that housed it. He was the lab's associate director after the war.

Manley was amazed that no one wanted to explain the operation of these simple physical devices. As I pieced together what I could of the process, he listened to my speculations, laughed at my diagrams, and gently corrected my errors.

A basic nuclear weapon—an A-bomb—is a fission bomb, like those dropped on Hiroshima and Nagasaki. They blow up when the nuclei of atoms in uranium (U-235) or plutonium (PU-239) are split rapidly. Manley waved his hands as he explained what is now elementary nuclear physics. At the time, fission was a barely revealed mystery.

Studying for his physics Ph.D. in the early 1930s, Manley had an old-fashioned skeptical teacher who thought Einstein wrong about time and space. Only in 1939 was fission explained, by Lise Meitner, a German Jew, and her nephew Otto Frisch. When Manley went to Los Alamos in 1943, many important details of the process were poorly understood. But the basics were known.

In a fission reaction, an atom of a complicated heavy element like plutonium is split into two radioactive nuclei, called fission products, and two or three neutrons. At the same time, a lot of energy is released. Get many atoms to do the same thing, in an uncontrolled fast chain reaction, and you get an explosion. The

---

that Exploded (New York: Random House, 1981). People at the weapons laboratories are forbidden to comment on Morland's book or his description of the bomb configuration, although one grinned and inclined his head in assent when I asked if Morland had given a good basic description. Lab scientists invariably refer the curious to a technical book not easily comprehended by the layperson, Samuel Glasstone, ed., *The Effects of Nuclear Weapons*, rev. ed. (Washington: U.S. Atomic Energy Commission, 1962, with changes as of 1964). See also the epilogue in Rhodes, *The Making of the Atomic Bomb*, esp. pp. 773–5; and John McPhee, *The Curve of Binding Energy* (New York: Random House, Ballantine Books, 1975).

best way to do that is to compress the fissile material into a small, dense mass so the atoms are close together. Then a blast of neutrons will break up some of the atoms and propagate more neutrons to attack more atoms. The Los Alamos scientists called the bomb "the Gadget" for secrecy's sake, and it is still just a gadget, albeit one that inspires more emotion than the standard kitchen appliance.

A-bombs no longer make up much of the superpower arsenals. They have been superseded by hydrogen, or H-bombs, which are much more powerful. But A-bombs are the tactical nuclear weapons NATO soldiers would carry in backpacks if the West decided to fight a limited nuclear war, and it takes an A-bomb to make an H-bomb go off. The weapons designers call the A-bomb at the heart of an H-bomb a "primary." In 1989, in the wake of scandals about the safety of the DOE's plutonium processing plant near Denver, the press began referring to the products of Rocky Flats as "plutonium triggers" for hydrogen bombs. Manley was tickled by the misnomer. The term was novel, and apparently invented by some public relations whiz. The innocent-sounding "trigger" for a thermonuclear weapon is itself a bomb.

The scientists from the wartime Manhattan Project produced two different A-bomb designs. The bomb dropped at Hiroshima was a "gun-type" weapon. Upon being triggered, a cylinder of uranium was shot into a larger target mass of uranium, plugging a hole that had been machined into the target. A supercritical mass was created, and the fast chain reaction released enough energy to almost wholly destroy Hiroshima.

The bomb dropped on Hiroshima was relatively simple. The Los Alamos scientists did not even bother to test a prototype of the gun-type bomb before loading it on the *Enola Gay*. They had complete confidence in their handiwork. Fat Man, used on Nagasaki on August 9, had a different and more challenging design that presented those infamously "technically sweet" problems. Fat Man was made of plutonium in a configuration still used in the primaries that make a modern H-bomb explode. The plutonium weapon was more problematic, which is why the Nagasaki-type bomb was tested on July 16, 1945, in the desert near Alamagordo,

New Mexico. The agreed-on code for a successful Trinity Test was "Baby Boy," and the baby *was* a boy.[2]

In these plutonium weapons, the fissile material is shaped like a hollow sphere, which gave Fat Man a shape like Churchill's. In the center of the original Trinity and Nagasaki bombs was an "initiator," a neutron generator made up of a small amount of the elements beryllium and polonium. A set of lens-shaped conventional explosive charges completely surrounded the plutonium sphere. Triggering these charges compressed the plutonium into a dense supercritical mass, allowing fission to take place. The compression of the plutonium also crushed the initiator in its core, causing the initiator to generate neutrons and perpetuate the chain reaction. Because the design requires compacting the fissile material into a smaller mass, these are called "implosion weapons."

The parts are a little different these days, but the process is the same. A typical A-bomb is a straightforward process. First a fuze is activated. A variety of fuzes are designed by Sandia engineers. The type of fuze that ends up in any particular bomb is determined by the structure of the bomb's delivery system.

An electrical detonator, also designed at Sandia, is triggered by the fuze, setting off the chemical explosive surrounding a mass of plutonium. The fuze simultaneously activates a neutron generator. I saw several unclassified neutron generators on a physicist's desk at Sandia Labs. Glass and metal tubes filled with lacy arrangements of glass and metal, the whole no longer than the width of a man's palm, they bear a superficial resemblence to old-fashioned vacuum tubes.

The chemical explosive compresses the PU-239 into a supercritical mass. The size and shape of the plutonium core is determined by Los Alamos primary designers with the aid of gigantic computer programs. The plutonium itself is made in government

---

[2] For these and other euphemisms see Spencer R. Weart, *Nuclear Fear: A History of Images* (Cambridge: Harvard University Press, 1988), p. 102; and Carol Cohn, "Slick 'ems, Glick 'ems, Christmas Trees, and Cookie Cutters: Nuclear Language and How We Learned to Love the Bomb," *Bulletin of the Atomic Scientists*, June 1987, pp. 17–24.

reactors, such as those at Hanford, Washington.[3] It may also be separated as a by-product of the fission in nuclear power plants, a concern to those worried about nuclear proliferation and the reason Israeli bombers destroyed Iraq's first and only nuclear power plant. The safe transport of plutonium and other elements for making nuclear bombs to Los Alamos and elsewhere in the Department of Energy's nuclear weapons complex depends on the Sandia engineers who specialize in the development of impervious containers and foils for potential thieves.

Once at Los Alamos, the plutonium is processed by the Material Sciences Technology Division, where Gene the technician spent three-and-a-half years as a materials handler. Plutonium feels warm to the touch. The plutonium used in reactor fuel cells is naturally hotter than the isotope that is used in bombs, but the temperature really depends on the size of the mass and its alloy. The heat is a result of the natural fissioning of plutonium nuclei. Some pieces are body temperature. Others can be hot enough to burn a hole right through a worker's glove.

As the plutonium is compressed in the bomb the neutron generator introduces a burst of neutrons into the supercritical assembly, beginning the explosive fission reaction. The plutonium is contained by a "tamper," or sheath, that both reflects neutrons back into the assembly and slows expansion from the reaction, thus sustaining the amount of plutonium available for further reaction.

The scientists who designed the implosion bomb held their breath when it was first tested. Manley remembered an air of great tension, tolerable only because there was so much work to be done.

The ignition of the lens-shaped charges of conventional explosives surrounding the plutonium core had to be perfectly timed. If they exploded off-schedule or failed to compress the pluto-

---

[3] For a description of the Hanford community see Paul Loeb, *Nuclear Culture: Living and Working in the World's Largest Atomic Complex* (New York: Coward, McCann, & Geohegan, Inc., 1982).

nium into a supercritical configuration, the plutonium would be blown apart before fission was complete. The precious substance, refined at great expense in the extensive and top-secret facilities at Hanford, would be scattered all over the New Mexico desert.

The risk of losing the plutonium seemed too great, so the scientists planned for the worst. They thought to explode the Trinity bomb inside an enormous steel bottle nicknamed "Jumbo." In the event of an unsuccessful test, the plutonium might be contained and recovered for further use.

Then people began thinking about what a successful test might mean: 214 tons of radioactive steel shards raining down, radioactive steel dust carried aloft on the desert winds. Jumbo was abandoned. Years later the army tried to blow it up with conventional explosives. The thick gray bottle was damaged but not destroyed. Manley thought their attempt demonstrated a shameful lack of respect for history.

John Manley is the historical conscience of Los Alamos. He was one of the first scientists to arrive in Los Alamos, back before the town was a town. Los Alamos National Laboratory has changed a lot from the old days when it was Los Alamos *Scientific* Laboratory and everyone worked together in devotion to a single common cause.

The librarian at the laboratory's Oppenheimer Memorial Study Center has an evolutionary flowchart describing the fragmentation of old divisions into new. The most famous old divisions are intact: though: the T-division (for Theory) and P-division (for Physics). Some who work in those divisions help with the design of modern nuclear weapons. But today the design work is done mostly by scientists and engineers in the Applied Theoretical Physics Division. Its nickname is X-division. They call their work "nasty physics."

No one seems to know for sure why the weapons division is called "X." One theory is that the name was shortened from the original designation GMX, meaning "Gadget, Mass, and Explosive." There is a commonly cited alternative explanation. "X is the unknown value in an equation, the thing you are trying to solve for," one physicist noted wryly. X is the mystery.

•   •   •

The eighteen-inch-high projected image of the cloud started to rise and change color. White gave way to gold and red, as the initial high heat dissipated and shifted down the color spectrum. The column shot up fast, its irregular round head extruding a ragged tail. The reds and yellows were joined by black, first rising up the stalk and edging the ruffles of the head, then dirtying the entire cloud. Some people saw a mushroom, like the poisonous white-spotted red fly amanita.[4] Manley suggested a black rose, its petals unfolding as it grew skyward. "It's not very nice," he said quietly, "but the black there is the city of Hiroshima."

•   •   •

Karl made a little self-deprecating joke as he struggled to adjust his overweight middle-aged frame behind a corner table in the Los Alamos Lab's public cafeteria. But he quickly got down to business. "These bombs are unrecognizable," he declared. "They're so much more subtle, so much more beautiful than they were in the old days."

Karl studied physics in graduate school in the mid-1960s. There he heard a lot of "dangerous talk" among the graduate students, talk so critical of the United States it seemed unpatriotic. He deliberately planned to make his career at the weapons lab.

"I wanted to make a contribution to national defense because it was clear to me that those folks at my school, and by extension at Harvard and MIT and other excellent schools, weren't going to do it," he recalled. "You remember there was a convulsion in this country during my formative years. In 1957 *Sputnik* went up. Admiral Rickover came on TV and the impression I got was that every

[4]Weart, *Nuclear Fear*, pp. 402–3.

red-blooded American boy that could had an obligation to be a scientist or an engineer." Now a group leader in Los Alamos's X-division, Karl never changed his mind. "Our survival depends on it," he said flatly, trying to make himself comfortable in the too-small chair.

Karl is a master of thermonuclear weapons design. In one of the indirect ways scientists at Los Alamos take credit for bomb design, Karl boasted of "nine Nevada shots with my name on them." The Nevada shots are weapons tests, conducted underground at the Nevada Test Site near enough to Las Vegas that gamblers sometimes feel the earth shake. With his name on them, these shots were his responsibility—his bombs.

That responsibility brings stress, the source of Karl's major complaint. Stress makes the turnover in the X-division groups way too high. He could always get replacements for the burned-out designers, he said, dropping the paper napkin he'd been twisting and holding his thumb and index finger four inches apart to illustrate the stack of applications to his group. But rapid turnover decreases institutional memory and slows the work.

His own stress was exacerbated when he began supervising the forty or so other designers in his group. Karl's sense of humor was so dry, his voice so deadpan, that frequently only he detected the joke that had prompted his periodic bursts of laughter. "There's constant pressure," he said by way of explanation. Laughter relieves it. Music helps, too. Karl gives French horn lessons, and in talking about them revealed an eerie talent. He has a fantastic and apparently unerring memory for numbers. For example, he knew exactly how many music lessons he had given (1,955). Other numbers flew across the table like precise little birds whenever he opened his mouth. They stay in his mind.

• • •

The subtle and beautiful bombs Karl designs are thermonuclear weapons, H-bombs. They involve a fusion reaction and are somewhat more complicated than the atomic bombs John Manley helped develop. In theory, they are unlimited in explosive power. The stars burn with thermonuclear fury, consuming isotopes of hydrogen, the simplest, lightest, and most common element in the universe.

In a fusion reaction the heavy hydrogen isotopes deuterium and tritium are fused together. The process creates helium, a heavier element, and releases neutrons and energy. Thus the stars shine. At Sandia, engineers who work on weaponizing the physics packages for H-bombs joke about their errand-boy job—"delivering buckets of sunshine for the taxpayers."

To make deuterium and tritium fuse, the mutual repulsion of the positively charged nuclei must be overcome. This means heating these components to millions of degrees. Only something like the heat of a fission reaction, the heat that can fuse glass or burn the shadow of a person into cement, is intense enough to cause fusion. That explains why a basic H-bomb has two stages. The primary stage is a fission bomb. The secondary, the part filled with hydrogen isotopes, is designed to produce a fusion reaction.

If you stop at that you get a "neutron bomb," because the neutrons from the fusion reaction are free to fly into the atmosphere. Neutron bombs provoked moral outrage in the early 1980s when the public learned of proposals to add them to America's arsenal. Emitting high quantities of deadly neutrons with relatively small explosions, neutron bombs can kill living things without destroying the physical infrastructure of a city. "They're like rifles," one scientist said, puzzled by the moral distinction between a neutron bomb and an ordinary H-bomb. You would not use a rifle to pulverize a city, but we do not condemn rifles. Why is a neutron bomb ethically any different?, he wondered.

An ordinary H-bomb captures those lethal neutrons and uses them to increase the explosive power and radioactive yield of the weapon. The secondary stage of the typical thermonuclear bomb is surrounded by a shell of U-238. This blanket of uranium acts as a tamper, holding the central mass together and reflecting some of the neutrons back into the core, where they encourage more fusion. The uranium shell around a thermonuclear weapon also absorbs some of the high-energy neutrons from the fusion reaction it encloses, undergoes fission, and thus releases more energy. A typical H-bomb is a fission-fusion-fission gadget, with about half of its explosive force coming from the second-stage fusion reaction and half from the subsequent fissioning of the uranium blanket that surrounds it.

The first thermonuclear device was exploded on November 1, 1952, on the Eniwetok Atoll in the Pacific. It was made hurriedly, in response to the Soviets' test of their own first A-bomb in 1949. Oppenheimer and other scientists on the General Advisory Committee of the Atomic Energy Committee advised against it. They saw no need to rush the development of a bomb promising to be a thousand times more powerful than the Hiroshima bomb. World history seemed at a crossroads, and they hoped diplomatic efforts to control warfare might be taken more seriously in the light of the two destroyed Japanese cities and the threat of bombs that could do worse.

Others, including Edward Teller, disagreed. Even during the war years at Los Alamos, when his colleagues were working hard on the simple A-bomb, he focused his energy on the theory of the H-bomb. The proponents of the accelerated H-bomb research program won Truman's ear. Along with Stan Ulam, Teller figured out how it could be done. He was known as the "Father of the H-bomb" and has barely disclaimed the honor.[5]

In the first thermonuclear experiments the isotopes of hydrogen were in a gaseous or liquid state. A model of the Mark 17, the first droppable hydrogen bomb, is in the Atomic Museum on Kirtland Air Force Base. Huge and grotesquely heavy, it is more like a thermonuclear factory hidden in an oval tank than a usable bomb. The Mark 17, part of the American arsenal between 1954 and 1957, weighed twenty-one tons. More efficient modern H-bombs are much smaller. They use hydrogen in a denser, solid form, like lithium deuteride or lithium hydride compounds.

The total fusion reaction results from two sources of energy. The fission reaction in the primary of a thermonuclear weapon pro-

---

[5] See Edward Teller, "The Work of Many People," *Science*, no. 122 (1955): 267. See also the comments in the epilogue in Rhodes, *The Making of the Atomic Bomb*, esp. p. 773.

duces energy in the form of X rays, or photons, that heat the hydrogen isotopes in the secondary. These photons can be "channeled," by mechanisms including reflection, so they enter the hydrogen mass from many angles. In a typical H-bomb, the secondary is encased in a plastic material that allows the free flow of neutrons and photons around it. Apparently polystyrene—Styrofoam—is a preferred material for this purpose. Besides keeping ice cold and coffee hot, it is light and quite transparent to high-energy radiation.

As the fusion reaction begins, the helium nuclei (also called alpha particles) that are created have a lot of energy of movement. They help to heat up the remaining unfused particles in the secondary, thus encouraging more fusion. In essence, therefore, a thermonuclear explosion is a four-stage process of unbelievable speed—a millionth of a second. First, a primary or fission bomb is detonated. In the secondary stage, a nearby mass of hydrogen isotopes is bombarded by photons from the fission explosion. Some enter by a direct route, while others are channeled through the plastic surrounding the mass and enter from other angles. The details of this process affect the shape and size of the explosion that results. It is one of the subtleties to which Karl alluded.

As fusion begins, the alpha particles from the fusion process further heat the mass and encourage more fusion. And the fusion produces free neutrons, which are released into the uranium shell surrounding the secondary assembly. Some are reflected back, some are absorbed by the uranium and produce an additional fission reaction, others go into the atmosphere—or, assuming this is a test and not an attack, into the rock in the deep vertical wells and horizontal tunnels at the Nevada Test Site, where Sandia and Los Alamos scientists measure the results with testing instruments so refined they can transmit their information to observers on the surface in the split-second before they are destroyed by the blast.

• • •

"I'd love to see an aboveground test," Karl said, drawing doodles with his plastic coffee stirrer in a rapidly drying brown puddle on the plastic tabletop. All but underground nuclear bomb tests were forbidden by treaty in 1963. The 1976 Threshold Test Ban Treaty went into effect just after Karl tested his biggest bomb, a 335-kiloton warhead. Nothing over 150 kilotons can be detonated under the terms of the treaty, and so Karl, in his early forties, was the Old Man in his group, with an experience none will ever claim as long as the treaty holds. But the big underground test did not satisfy Karl. He wanted to experience the bomb unmediated by instruments and unmuffled by rock.

"I've seen the films. I want to feel the heat. This is the most there is; this is the closest you get to playing God," he said, in the same strained flat tone he'd used to describe his love for music and for teaching young scientists how to design new bombs. "You know," he continued, "stars and bombs are a lot alike, but you don't get to design a star and see if it'll go supernova when you want it to." He does not expect a war. "But," he affirmed, "if those megatons roll, this country is going to have the best."

●　　●　　●

A young engineer at Sandia was talking about his work. "I've been depressed for weeks on end," he said between jokes. He does not work on bombs.

At first he was assigned to a Sandia group devoted to nuclear safeguards, working on a project intended to prevent sabotage of nuclear power plants. Finding the work lacked challenges, he transferred to another group, this one responsible for predicting the consequences of nuclear reactor accidents. "I did consequence analysis—also known, if you want to be a jerk about it, as 'ex-plant analysis,' which is modeling the atmospheric transport of crap, to use a nontechnical term, from a nuclear reactor." That project seemed dumb, too, since no one had applied for a nuclear reactor license for years. He looked around the lab for more interesting and relevant work.

This restlessness is common at both Sandia and Los Alamos. Karl's burned-out weapons designers in Los Alamos Lab's X-division rarely leave the lab when they leave his group. If they can't stand the pressure they transfer to another division, using the internal help-wanted ads that both labs list for their employees.

The young man continued. "I applied for a transfer about six months ago. It was actually a really dream job in some aspects, and the guy offered me the job." He hesitated. He had concluded that his research in that organization would be applied to bombs. He refused the job. The disappointed supervisor agreed to forget he had ever seen the young man's application.

Why did he turn down a dream job? "I didn't like the atmosphere," he explained. "It's a bomb factory." In the new position he would have had to work with bombheads. And bombheads are notorious. They get all caught up in what they falsely imagine to be fascinating details of bomb design. They identify with the bomb. They *like* it. They reinforce each other in their devotion to the bomb. For him, turning down a bomb job was a moral accomplishment. Saying no momentarily relieved what he thought was "existential ennui."

At Los Alamos, weapons designers are not called bombheads. The weapons designers themselves are in charge of the vocabulary, and they say they belong to a "club" with its own rules and norms—the "weapons design culture."

Karl's Ph.D. thesis adviser had warned him not to go to Los Alamos. "He said it gets boring," Karl said, twisting the plastic coffee stirrer around his index finger. "It doesn't. I love it. It's—it's the thrill of the chase." The fox before the hounds is a subtle, beautiful, and innovative design that works. For example, can you double the yield of a warhead without an appreciable increase in weight? The chase involves computer codes that are 100,000 lines long. That's 100,000 individual instructions that take 100 hours to run on Los Alamos's Cray computers, the most powerful in the world.

In the old days, Karl said, referring to the 1970s, those computer instructions were all punched on cards. Designers would take a carefully arranged deck of computer cards down to the Central Computing Facility and hand it to a technician. Later they'd retrieve the cards and a bundle of printouts and go back to the office to see what would happen if *this* little variable were changed or *that* little relationship altered. Today the bomb designers all have terminals at their desks, although they still double up in the offices behind the new security gate blocking access to the lab's administration building.

"I kind of feel sorry for the people who think design work is sterile," Karl said. Karl meant it when he said he loved his job, frustrations and all. He was clearly not in it for the money. In the mid-1980s, new Ph.D.'s in his group earned annual salaries of about $43,000. Senior people with administrative responsibilities like Karl made not much more—he said around $50,000 a year. But income from an inheritance was enough to keep Karl's family comfortable forever. He used his salary for fun on the stock market.

Weapons designers contend their work is fascinating, but laboratory professionals more distanced from the design work repeat the charges incessantly: design work is dull, sterile, and scientifically insignificant. John Manley always figured it to be about as exciting as designing a new toothbrush, more a matter of engineering refinements than a scientific challenge. Once you understood the basic configurations for atomic and thermonuclear bombs, where was the challenge in minor tinkering? The development of all these new designs for toothbrushes, Manley admitted, is cause for chagrin, but not retreat. He still thinks modern weapons design is a waste of scientific talent.

The conventional wisdom outside the weapons groups at Los Alamos and Sandia sneers at Karl's love for his projects and contradicts the official rhetoric of the laboratories, in which basic scientific research is important but never more than a close second to their weapons design work. One might imagine this would automatically confer higher status on the designers in the weapons groups at the labs. Not true. At both laboratories, but especially in Los Alamos, their coworkers show no sign of envy. The stack of applications on Karl's desk represents a small fraction of Los Alamos scientists.

Thus the nuclear weapons club, with its own culture, is relatively small. What are the values of that culture? Designers must be good at the details and find them engrossing. That is a requirement for membership in the club. Once admitted, a scientist must understand and respect the cardinal rule of bomb designers: Don't Slip a Shot. That means do not get so far behind on your work that a planned demonstration at the Nevada Test Site might be delayed. Stick to the schedule. *This* is the pressure that sends designers rummaging through the laboratory's postings in search of a new position. Weapons design work is hard because the engineering details are complex, but it is stressful because the bureaucratic deadlines are unrelenting. Modern designers voluntarily impose on themselves an artificial sense of urgency that parrots the response of the original Manhattan Project scientists to the horrors of war.

And beyond the imperative "Don't Slip a Shot"? What else defines the weapons design culture? Behind the common commitment to meeting deadlines lies nothing of substance. The weapons design club holds no meetings, keeps no minutes. Every weapons designer I asked emphasized the importance of their culture but could think of nothing to define it besides the emphasis on punctuality and responsible work habits. Just as a business corporation is a legal fiction pretending to be as real as a person, the weapons design culture is a moral fiction imagining itself to be a real culture. Weapons designers' bragging about their "club" seemed simply an attempt to deflect outsiders' attacks on their "dull" work, their misunderstood love.

•   •   •

Sometime in the past, Agnew's film of the destruction of Hiroshima was mercilessly cut and spliced. A security angle may have dictated the editing. Displaying the entire course of the explosion in real time would allow clever adversaries to know more about the bomb than U.S. policymakers might wish. Supposedly even more recent H-bomb films are altered to disguise the expansion rates of the smooth and glossy smoke-colored fireballs. Some bit is always left out, or the perspective changed, so the process is never completely coherent.

In Manley's copy of Agnew's film, the original process of the explosion was chopped into bits, and each bit was copied and shown several times. A single event through time thus became a series of short repetitive scenes. Projected on the wall, the white cloud starts to grow. It is halted in its expansion, and the scene begins again. The white cloud starts to grow. Again it is arrested in its development. Eventually the scene shifts. Now the fireball starts to ascend. It rises for a few seconds, then stops. The scene repeats.

The whole film is only a few minutes long, but time changes character with the edited shifts and breaks and Agnew's unsteady hand in the plane buffeted by the explosion's shock wave. In the fragments of the film, time does not flow. This technique is now deliberately used by filmmakers to signify a departure from normalcy, a hallucinogenic state, dream, or nightmare. The black stuff that used to be the wood, paper, and people of Hiroshima suffuses the cloud, retreats, repeats, retreats, repeats.

After the *Enola Gay* delivered the bomb, the pilot of the instrument plane flew a slow semicircle around Hiroshima so the observers could get a full view of the blast. In the rocky cadence of the film, a surprising horizon of bright blue appears behind the smoky column. It establishes perspective and releases the viewer from the heart of the cloud. Over the clicking of the projector, Manley said, "That's the Inland Sea." Then the image disappeared. The movie was over.

5

EVERYTHING IS
UNDER CONTROL

.
.
.

Gene was a technician at Los Alamos National Laboratory for six years. More than half of that time he spent in the lab's plutonium processing facility, where the precious material is worked and reworked, separated from its old alloy and reshaped for new purposes. Gene's job was reprocessing the radioactive materials from old nuclear weapons so they could be reused. He handled the guts of nuclear warheads.

When asked to describe how that feels, he stared blankly, shook his head, made a few false starts, and finally, his face brightening, settled on an analogy.

"Holding a bomb is like pointing a gun at somebody," he said. "You have to do it to understand the feeling of raw power. It's *awesome* to hold a bomb that powerful."

Most people have never held a gun on someone. Imagining the effect of even that experience is no easy task. When children point sticks or fingers and yell "Bang!" their victims can yell back, die dramatically, or refuse to die at all ("You *missed* me!"). Imagine instead controlling a weapon that can turn a city to ash.

Away from the Hill, Gene thought of the bomb, abstractly, as "a sin we have to learn to live with." But as a technician, the bomb was a piece of material to be handled with great care using detailed Standard Operating Procedures (SOPs). Gene worked with the bomb inside a sealed glove box. He was insulated by layers of gloves taped tight to his sleeves and protective clothing studded with radiation monitors. "We have data sheets," Gene's former MST supervisor told me. "They tell you pretty much exactly what the steps are, what to do. You follow that, you can't go wrong." To make sure, they had one supervisor for every five technicians.

Sometimes the technicians were careless, the predictions about what would happen inside the glove box were wrong, the SOPs inadequate for controlling the process, and the supervisors out of sight or out of their league. And SOPs can't overcome opposing natural impulses. Gene saw a technician seriously burned when someone accidently opened a valve on a pressurized cannister of acid. Dreading the cheers and catcalls of her male coworkers, she, like some other women in the plutonium processing facility, refused to strip. Gene's supervisor forced her out of her clothing and wondered later if he had done the right thing.

Gene had learned about radioactive elements from his father, a Los Alamos physicist who had helped write the protocols for dealing with radioactive materials. Gene was assigned to work with americium, a particularly toxic and volatile element, because he was smart and careful. Even so, he had exceeded the recommended radiation exposures between fifteen and twenty times. He kept meaning to put a card in his wallet, in case of emergency, requesting no unnecessary X rays.

To Karl the bomb was a small sun going supernova by design. The intricate beauty of the details and the glory of controlling the process were marred only by the test ban treaties, human inventions that kept him from feeling the heat he imagined would hugely and objectively mirror his own subjective radiant flush of power. To the Sandia engineer with existential ennui the bomb was a moral challenge. To John Manley and the first generation of atomic scientists, it was both a moral necessity and a technical puzzle.

At the Trinity Test there were three scientific observation posts. As the scientist in charge of blast and shock measurements, Manley supervised the one ten thousand feet west of ground zero. The narrow bunker was crowded with equipment, he remembered, and nervous men. In T-shirts and shorts, they hovered over their instruments. A periscope-type projector was focused on the gadget so they could safely watch the explosion. But the westward-facing door of the bunker was open, and when the bomb went off Manley forgot to watch the small image projected on the wall beside the door. "The light!" he said forty years later. He was stunned by it. Suddenly everything outside was bleached white.

They had predicted success, of course, but there was still uncertainty, a tantalizing chance of failure. When Manley saw the light he felt tremendous relief. The unnatural glare meant that the wizards' work on the Hill had paid off. Then he turned back to his instruments.

Manley thought they didn't do science on the Hill anymore. I heard him say it on many occasions, always with a little laugh. But this man tells few jokes that do not advance his incessant pedagogical purposes. When he said there are no true scientists at Los Alamos National Laboratory, he meant something serious.

"I guess I'm being stuffy, in the sense that I'm making a distinction between a true scientist, in terms of motivation, and these people who work in technical fields," he explained one day. There was nothing wrong with being a technician, a technologist, or an engineer. Manley himself began as a student of engineering, motivated by a high school infatuation with *Popular Mechanics*. From it he got instructions on how to build the newest scientific gadget—a crystal radio set—and an insatiable appetite for more. As a college student, however, Manley quickly tired of the "cookbook approach" in engineering. Do you want to bake a cake? Follow these instructions. Do you want to build a bridge, or a radio set? There are recipes for those, too. But you will never know *how* they work, why the cake is a cake and not something else.

For the scientists at Los Alamos, he thought, research had become routinized. Tied to the imperatives of the military, it was only a shadow of what it would be were the government's "programmatic" constraints removed. True scientists seek truth, not better toothbrushes.

That may seem harsh, Manley conceded, but so be it. He proposed a simple test. If someone claiming to be a scientist concerns himself with the moral and political significance of his work—at a weapons lab, that means the arms race—he passes the test. If he slights the issue, he fails as a scientist.

Choosing his words with care, Manley said, "My impression—and that's all it is, of course—is that there are far too few scientists now, and let's distinguish them from engineers and technicians, who are really and genuinely deeply concerned about these questions. Well, the list is almost clear. You can take the sponsors of the Federation of American Scientists or the Union of Concerned Scientists—and there you have it."

We were in the Manleys' living room in Los Alamos. It was a warm day in early spring. The reflected sunlight was diffused by the polished wooden floor and his wavy white hair. He looked at the tape recorder on the coffee table, then out into the yard where Kay, his wife of more than fifty years, a fervent feminist and an avid gardener, struggled to place a lawn sprinkler in just the right place.

"Let me just be unfair in a way, probably, by saying that it seems to me that the most responsible and deeply—I'll even use the word—*human* scientists that I know of are the ones who are concerned about these questions," he said deliberately. "And there are many who are not. Those who don't consider such questions are not people I like very well."

He stopped. Did I understand? No.

"The reason I don't like them, you see, is because I don't consider that they're even being good *scientists*. Now why do I say that?" he asked rhetorically. "I think that there are certain messages in the scientific method, in the reliance on logic and discussion and all that sort of stuff, that are generally valuable to humanity."

Scientists who neglect the moral and political consequences of their work made him sad. "How can they be sufficiently curious about the world in which they live to be a good scientist," he wondered, "and not raise these questions?"

Every six weeks or so Manley takes his modified Dodge van into the wilderness. There, sitting in a lawn chair in the shade with the portable typewriter he won as a prize in college, he writes letters, essays, and reviews of books on what he calls "the nuclear arms predicament."

Lately he has been obsessed with the seeming repetition of history. In 1949 the Russians exploded Joe, their own atomic bomb. It sent American policymakers into a frenzy of confusion. The United States' powerful special invention had been mastered by a new enemy, and no one knew what to do about it. Manley was the executive secretary to the General Advisory Committee of the Atomic Energy Commission. The General Advisory Committee, headed by Oppenheimer, was asked what it thought of the idea of setting up a crash program to develop a "superbomb," the H-bomb. The scientists on the GAC deliberated, called witnesses, argued among themselves, and recommended against an accelerated H-bomb program. They reasoned that it might begin a nuclear arms race.

After observing their proceedings and doing research for the committee on the possible military uses of a bomb that promised to be "a thousand times as powerful as Hiroshima," Manley was impressed with the wisdom of their advice. President Truman, with conflicting opinions before him, was not. America's development of the H-bomb began the nuclear arms race. In Manley's view, a bad idea became public policy much too easily. The Strategic Defense Initiative reminded him of the H-bomb decision.[1]

---

[1] For the classic argument that the H-bomb decision began the arms race see Herbert F. York, "The Debate Over the Hydrogen Bomb," *Scientific American*, Oct. 1975, reprinted in *Arms Control and the Arms Race: Readings from "Scientific American"* (New York: W. H. Freeman, 1985), pp. 21–28.

"It's a replay of a very bad movie," he said. The H-bomb decision had ignored the advice of technically sophisticated advisers, just as Reagan had bypassed his science advisers when he announced his pursuit of a space shield to make nuclear weapons obsolete. Manley thought neither decision showed much concern for long-term international consequences. Why, he wondered, do people refuse to learn from experience?

Manley speaks like a scientist, writes like a scientist, and reasons like a scientist. Every idea must be clearly linked to its antecedents by explicit and valid logic. Every claim must be backed by evidence and a description of how that evidence was obtained. Every opinion must be formed fairly, on the basis of evidence and logic, imagination constrained by truth. Having built his case for excluding most of Los Alamos from the brotherhood of science, Manley felt compelled to concede that the moral failure was collective. He found young physicists to be poorly educated. They learned the method of science, but not its spirit.

That spirit he thought best expressed in the credo of the British Royal Society, dedicated to the exploration of nature "for the greater glory of God and Man." If science is not for that, he believed, it is hollow.

This is a vision from the old world. Understanding natural law is an end in itself, an act of love and a form of worship. In the new world, the modern world, science (and nearly everything else) is a fuel. It incites change, movement, and competition. It is an instrument for mastery and control, but also domination and repression.[2]

• • •

"The status! Oh, the status is wonderful," Gus crowed. We were in a cheap diner drinking strong, bitter coffee, and Gus was explaining his feelings about working at Sandia.

---

[2]The point is well made in Marshall Berman, *All That Is Solid Melts into Air: The Experience of Modernity* (New York: Penguin Books, 1988).

The son of a Greek immigrant, he felt that his master's degree in electrical engineering had taken him far beyond his working-class family's wildest dreams.

"My father was beaming with pride that his son could work in a national laboratory," Gus said. Los Alamos or Lawrence Livermore would be more distinguished, he added quickly, but Sandia was definitely worth writing home about. "My cousins—they think we're doing magic," he laughed. "They think I have some special knowledge of world events, and especially the youngsters think I have secret knowledge of weapons."

After nine years of weapons work, Gus had transferred into a solar energy group. There he did unclassified research on photovoltaic cells. But the misty mantle from his weapons work days still clung to his shoulders. Sometimes it made him uncomfortable. Waving his hands before an invisible orchestra he said, "There's this mystique about it. I don't know how to dispel it." The more he explained things to people rationally, the more they responded with awe and wonder.

Although Gus was not, strictly speaking, a scientist, he understood, believed in, and practiced the scientific method. He naturally used it to fulfill his civic duty. Alarmed by too many cars racing down a residential street filled with children and pets, Gus decided to convince the city to reroute local traffic.

We faced a similar problem in my own neighborhood a few years ago. The sound of squealing brakes and breaking glass had become all too familiar. I made phone calls threatening the traffic department with a petition drive and publicity campaign and begging a friend in city government for relief. Two weeks later I walked out my door and discovered, to my shock, that the dangerous corner had four new stop signs.

Gus approached his neighborhood traffic problem rather differently. He designed and built a traffic counter in his garage workshop. With it he measured the flow of vehicles for several weeks. Gus then analyzed the results and prepared a detailed presentation that flabbergasted the authorities and convinced them to shift a chain of stop signs. His grateful neighbors insisted that he be appointed to a citizens' advisory board and elected him president of the neighborhood association. Then Gus made a new discovery.

"When I first started in this whole thing, I had a typical engineer's orientation—that everything was cut and dried," Gus said cheerfully. "And I was *shocked*. You learn in your seventh-grade civics class that there are these models of how government works, and then I found out it didn't work that way."

Faced with the limits of the engineering worldview, where every problem must be amenable to rational analysis and near-perfect control, Gus responded like—an engineer. "It's a tribute to my engineering background that what I attempted to do is change my model to fit the reality," he explained. "Being involved in politics, I understand that things are not always black and white, and there's more than one way to skin a cat." He'd learned that cat-skinning can itself be *engineered*.

The lessons of politics helped him at Sandia, too. The problem with Sandia, Gus charged, was all those scientists and engineers. Accustomed to clearly defined questions with relatively straightforward answers, they too rarely practiced the subtle art of diplomacy.

"With volatile issues, it's [a question of] how the truth is presented," he explained. What volatile issues? Gus beat around the bush for a while, then signaled the waitress for another cup of coffee.

It wasn't a crucial matter affecting national security, he said quickly. It was only a human relations problem. His group at Sandia was unhappy with one of the private firms that supplied some of their photovoltaic equipment. The managers there were inefficient and uncooperative. His Sandia supervisors wanted to cut off relations with the firm, but they needed a good excuse. Unreliable equipment would be reason enough. Unfortunately, the equipment was fine. Gus could ease the burdens of Sandia bureaucrats by modifying his scientific conclusions just the *slightest* bit. In other words, they wanted him to doctor his results.

"The supervisor comes in and says, 'I want to see what you're doing.' And then the [nit]picking starts. We might go on like that for months," Gus said. "The management knows what we're doing and they know you can manipulate the data, and they put the pressure on. Most of the time we try to stay objective." Once in a while, Gus admitted, he would compromise or surrender. His experience in local government had taught him when to cut his losses.

Politics also helped him understand a frustrating practice at Sandia. Before a technical report is published between Sandia's covers, it must make it through an internal obstacle course. "You need two peer reviews," Gus said, "then it goes up the hierarchy to the division supervisor, to the department manager, to the director, up to the vice-president." He ticked off the hurdles on his fingers.

Half of the time the red marks in the margins had nothing to do with technical content. "You get vice-presidents correcting your grammar," he said. "You ought to *see* what some of the comments are. Lately they've been looking for things that defame the laboratory or could cause problems with contractors."

The endless oversight caused perturbations among the staff who believed, as Gus once did, that when you discover the truth you just lay it out for others to judge. Your reputation depends on honesty, because the ultimate test is nature. Within a reasonable margin of error, your predictions are either correct or incorrect, your procedures and devices either fly or fall.

This is a universal tenet of the scientific and engineering communities. Good science and engineering means following the rules, exercising caution and restraint, and, above all, telling the truth. Gus and a few others told me they had been prodded to accept less rigorous standards. These complaints were unusual. Almost everyone at the weapons labs made it a point of pride that they were never asked to fudge their data or overstate its implications.

As for the political implications of his work, Gus had an answer to critics of the arms race. "I usually try to turn it around and say, 'Isn't suicide immoral?'"

"To most people, that'll usually take them aback, and they'll start thinking about it." The offensive strategy worked. "You can make a pretty convincing argument," he continued. His father, a blue-collar die-hard Democrat, would shudder if he ever learned the source of Gus's idea. "William F. Buckley—that's where I got that," Gus admitted. "I took the argument from him."

"When I was working in the weapons area," Gus went on, "I always felt that I was doing a direct service to the people, whether they appreciated it or not." That work was the primary function of a national laboratory, he reminded me, and objectively more important. But his unclassified solar energy research was more visible and popular with the public. "In ten years you might have a photovoltaic system powering your house," Gus said. "But you're not going to have a nuclear missile in your backyard." I cleared my throat. We were in a coffee shop three miles from Kirtland Air Force Base and Manzano Mountain. "Oh—well, ah," Gus said, "you know what I mean."

Gus laughed at his cousins for thinking him a magician, but he enjoyed the aura of power, too. Scientists and engineers have guaranteed themselves a place among the authority figures in Western cultures because they have succeeded so well in their work. More than any single group, scientists promise and deliver the power of transformation. Put simply, they are good at predicting and controlling nature.

Nuclear weapons turn matter into energy. Atoms get rearranged. Molecules fly apart. Things burn, sand becomes glass, and all on a massive scale. Harnessing these impressive but frightening powers requires the expenditure of public resources on a correspondingly massive scale. Thus there is public ambivalence about the ultimate worth, even in monetary terms, of science and its technological offshoots.

Ambivalence about science is nothing new. The imaginary Dr. Frankenstein wrecked the pastoral innocence of the countryside. His urban counterpart terrorized the civilized psyche with the irrepressibly primal Mr. Hyde. And the real Galileo exiled us from the center of God's heart when he argued for the heliocentric heresy. Each attacked superstition and ignorance and the accepted conventions about the human place in the scheme of things. All came to a bad end and each became a symbol of the tragic flaw of hubris. Their stories of unrestrained enthusiasm for scientific progress warn that even the most noble intentions can go awry. Ideas and technologies meant to be morally neutral can throw old and comfortable ways of thinking and acting into a tailspin. Threatened or on the verge of destruction, cultures strike back. Go too far, investigate too boldly or presume to control too much, and nature itself may exact punishment.

Modern science does not merely transform and destroy institutions, symbols, souls, and psyches. The whole of the real physical stuff that composes and sustains us can now be deliberately and completely rearranged so dramatically it seems unlikely that any human scientist would be left to analyze the remains. In light of that fact, does the social responsibility of scientists extend beyond integrity in research and careful quality control?

"I should keep myself informed," one man said. "If you can't look in the mirror when you shave in the morning, then quit," said another. "You have a responsibility to communicate what you know to the public," said a third. Many scientific professionals at the weapons laboratories felt their social responsibility, at least in principle, extended far beyond simple prescriptions for honest reporting of the results of careful laboratory procedures.

Karl, the master of thermonuclear design, had agreed. "What is your responsibility to history?" he asked, rephrasing the question.

In his flat, deliberate voice Karl marshaled evidence to suggest his answer. "I observe that in the years since the nuclear attack on Japan, there's been no world war in which millions of people have died. We've had wars, of course, and I grieve for those people, every one. But it's nothing compared to the fifty million that died in World War II." He began to catalog the brutality in human history. Looking backward, Karl saw generation upon generation of soldiers, miserable themselves, causing misery for others. Looking around, he saw instability and trouble in the Eastern bloc and a misguided antinuclear movement at home. So looking ahead to the "foreseeable future," he saw more weapons.[3]

---

[3] The estimate of fifty million dead in World War II is probably high. R. R. Palmer and Joel Colton, *A History of the Modern World*, 5th ed., (New York: Knopf, 1978), p. 819, put the figure at thirty-five to forty million.

"Someone who says they want to rid the world of nuclear weapons is saying, 'Let's make the world safe for conventional war,'" Karl said. His money, and more than that, was on deterrence. "My little boy will be five next month," Karl said, straightening his back, "and I don't want him to die in some war slogging through the mud."

Filled with victims of fierce cruelty in numbers he could not forget, Karl's reading of human history had convinced him that the risk of white-hot megatons raining on our heads was the unnegotiable price of averting the gruesome certainty of large-scale conventional warfare. Anyone looking at the facts, he thought, would understand that deterrence is a moral necessity.

If the facts of the world make certain actions necessary, then how do we know which facts matter most? How much danger are we in if we disarm? How much if we continue the arms race? The issue revolves around the quality of prediction, the very skill that scientists prize so much.

Prediction is easy if the problem is clear and the relevant natural principles known. Muddy the problem, multiply the variables, make realistic experimentation impossible, and the task becomes correspondingly more complex.

Several scientists chose DDT to illustrate the point. DDT was trumpeted as a great discovery. It promised only beneficial outcomes: malaria a thing of the past, increased crop yields, large victories for humans in their competition with insect species. Who could have predicted what would happen as the miracle chemical worked its way up through the food chain? And political variables are even more difficult to understand and predict, everyone agreed. Most thought social science a pitiful joke.

"We are the discoverers," one scientist said. "I'm not going to blame Christopher Columbus for all the problems in Latin America. How could he know where it was all going to lead to?" The wizards don't claim to have all the answers. Doing their honest best, they expect to be forgiven when their predictions turn out wrong.

Technicians at the laboratories are less sanguine about the question of control. The blue-sky guys on the professional staff seem to miss the obvious. "They're so intelligent in their field," one Los Alamos technician commented, "and they haven't a bit of common sense."

The most pervasive political concerns of lab employees revolve around organizational politics, the power struggles among co-workers and bosses. Sometimes these power struggles are trivial amusements, like the "space wars" battles over office and laboratory space. Sometimes they have more serious consequences.

"We used to call ourselves mushrooms," Gene said, speaking on behalf of his fellow technicians. "They keep us in the dark and feed us bullshit. In general, the staff members think they do it all and the technicians are just there for the ride."

The technicians resent this, of course, but they also find it dangerous. Their superiors, confident they understand the laws of nature, forget these also apply to themselves. They equate knowledge with control. So they leave hazardous chemicals lying around in glass beakers, snake electrical cords in twisted ropes across open doorways, set up huge banks of storage batteries next to open-flame boilers, throw tools around like children. "Seat belts!" one technician said bitterly. He'd waged a losing battle with the Los Alamos Lab's administration on behalf of a serious seat-belt safety campaign. With the lab covering so great an area, people drive constantly from tech area to tech area. Many believe themselves immune to the inertial force.

"They scare the hell out of me," another technician confided. After a long military career, Ernest earned $32,000 a year as an equipment maintenance supervisor at Los Alamos.

"I've got an impossible job. We've got over a thousand buildings in this laboratory," he sighed. Ernest was responsible for developing inventories and maintenance procedures for every piece of equipment in every building. Los Alamos Lab's management was unsympathetic to the magnitude of the task. "As soon as I had it going pretty well, they pulled one of my men off. This is an ongoing thing. You have to reinventory these buildings constantly. It's just maddening." Sometimes the inventories required on-site inspections. Often that meant trouble.

"I have to deal with scientists and group leaders," Ernest said.

"I'll go out and see their equipment and say, 'You're not supposed to do this.' In just a couple of minutes they let me know I'm not supposed to talk to them that way." He can't even influence, much less control them. "I'm just a high school graduate," he explained patiently. "I'm just a technician." Ernest found the lack of common sense, the disdain for someone who is *only* a high school graduate, and the infuriating intricacies of the bureaucracy profoundly frustrating. But he stays. "They bought me," he said evenly. "I can't quit now. It's too much money. I have four years, eleven months, and twenty-three days, give or take a few, until I can retire."

Ernest was paid to worry, even if the staff members just thanked him coldly and told him to go away. The battery rig by the open flame will probably not explode. People learn to step over the knotted cables on the floor. Responsible technicians fight to have plastic sleeves put over glass jars of nasty chemicals and resign themselves to picking up the tools at the end of the day.

Sometimes things go wrong, though, and the dire warnings of the sorcerer's assistants are vindicated through small disasters. Gene walked into a laboratory one day just as a staff member (with a Ph.D., Gene added) sliced three fingers off his hand with an electric saw. Thinking everything was under control, the fellow had not bothered to lower the safety shield over the blade. Gene remembered his Boy Scout first-aid lessons and laid the man down, tied his wrist, found the scattered fingers and bound them tightly to the palm. Then he ran for help. Later Gene received a written reprimand. According to Standard Operating Procedures, he should have gone for help *first*.

•　　•　　•

Grant that a bomb is like a gun. The effects of a gun depend on numerous variables. For what purpose was it designed? Is it well made, clean, and free of rust? Is it loaded? How skilled is the person who holds it? What is their intent? Is the target shielded or quick on its feet? Will they pull the trigger?

Smart people assume all weapons are loaded. But without a bullet in the chamber, a gun is only a psychological bluff, or at best the proverbial blunt object. Unlike a gun, however, a nuclear weapon contains parts that are, in and of themselves, deadly. To understand what could go wrong, begin at the beginning and imagine the worst.

People could try to steal weapons-grade uranium or plutonium and threaten to poison the countryside or build their own personal nuclear bomb. Someone could try to steal or sabotage an existing nuclear weapon while it is being moved from place to place or while it waits in stockpile, submarine, or silo. Those with access to nuclear weapons could go mad, imagine themselves promoted to greater authority, and assume the power to "send up the balloon," as they say. A nuclear weapon could go off accidentally, as its finely balanced components deteriorate with time or inconsiderate handling. And what if it simply falls out of an airplane?

In fact, one did, in 1957, and it landed on Albuquerque. A 42,000-pound Mark 17 thermonuclear bomb was inadvertently released from a B-36 bomber flying over the city. The first "deliverable" thermonuclear bomb was so heavy that pilots reported their planes rose hundreds of feet after it was dropped, "as if the bomb released the plane rather than the reverse." Had there been a nuclear explosion, it would have been on the order of one megaton. The conventional explosives inside the Mark 17 did detonate, creating a crater twelve feet deep and twenty-five feet in diameter and contaminating a field south of Kirtland Base with an unspecified amount of radioactivity. The accident was a well-kept secret until 1986, when David Morrissey, an investigative reporter for the *Albuquerque Journal*, hit pay dirt with a request for government documents under the Freedom of Information Act. His exposé fascinated the world press. Local residents greeted it with indifference.[4]

• • •

The gun analogy came to Gene as a flash of insight, but others at Sandia and Los Alamos had given it more thought. "I would feel

---

[4]The quote is from Carson Mark et al., "Weapon Design: We've Done a Lot but We Can't Say Much," *Los Alamos Science*, vol. 4, no. 7 (Winter/Spring 1983): 161. See David H. Morrissey, "Documents Explain Kirtland Accident: Safety Re-

uncomfortable about working for Remington firearms," Jerry said, talking between bites of Mexican food served by a harried waitress in a small family restaurant in Albuquerque. He had just finished a workout at his gym, and he was hungry.

A physicist in his mid-thirties, Jerry had left a teaching post at an Ivy League school for Los Alamos, where by his estimate nearly three-quarters of his work over a six-year period was on weapons projects. His motivation was simple. "Here I am, scraping by to buy a new oscilloscope, and the guys at Los Alamos had junked a hundred of them because they weren't the latest models." He packed his bags and moved to the top of the mesa.

Several weapons scientists claimed that working in a handgun factory was morally wrong. The effects of handguns were too well known, and Saturday night specials get *used*. Nuclear weapons are good for scaring enemies, but no one wants a nuclear war the way a desperate kid wants the fifty bucks from the liquor store cash register or a jilted lover, revenge.

Karl went to work at Los Alamos out of patriotism. He was not alone in his motivation. But some younger weapons lab employees think patriotism an unconvincing justification for weapons work, the easy answer to a hard question. Jerry mistrusted the political arguments.

"My strongest feeling is that we really have to maintain a strong defense in order to maintain our world position," he said initially. Then he started to evaluate what the United States had done with its position, and he wavered, finding a rough moral parity between the superpowers: a Nicaragua for every Afghanistan.

"There are a lot of questions I'd like to have answered," Jerry said abruptly. "What would be the social and political implications of our losing our world position? What would the Russians really do?" He shook his head. "They haven't really contracted, they al-

---

lease Mechanism Blamed in Dropping of 10-Megaton Weapon," *Albuquerque Journal*, Aug. 27, 1986, p. A-1; Doug Willis, "Former Crewman Describes H-Bomb Accident," (AP) *Albuquerque Journal*, Aug. 29, 1986, p. A-1; Jim Arnholz, "The Bomb? What Bomb? Oh, That Bomb . . . No Problem," *Albuquerque Journal*, Aug. 29, 1986, p. A-6; and Michael Perkins, "H-Bombing Navigator Faults B-36," *Albuquerque Journal*, Sept. 1, 1986, p. A-1.

ways seem to expand. We do the same thing, but . . ." He looked confused. I said, "But we're the *good* guys," and he laughed.

"But then there are the other questions," he said, turning serious again. "What *if* one of these things actually should be used? What *if* there actually was a global nuclear war? Could this actually ever happen?"

One day four hundred Los Alamos researchers, including Jerry, were called into a meeting and told that their research funding had "gone away." "They called us in on a Tuesday morning," he remembered, "and said, 'Well, what are you planning on doing next week?'" Everyone scrambled for the internal job postings. That was when Jerry decided to get out.

"See, at these national labs, when the group is not in the mainstream, when you're not 'intimately connected with the mission of the laboratory'—that's a phrase they bandied about a lot and nobody was ever really sure of what it meant—your funding is always in jeopardy," he explained. He called an old college buddy at Sandia. They found a place for him right away. He'd only been in Albuquerque for four months, but already he felt better about Sandia than he'd ever felt about Los Alamos.

Jerry had a hard time specifying how much of his Sandia research on radiation-hardened microchips was weapons related.

"They're used in weapons, they're used in communications, particularly communications satellites—and that's about all that I know," Jerry said. At Los Alamos, he thought, there was a clearer distinction between weapons programs and basic research, so he could easily identify the purpose of his projects. Sandia was more frankly an ordinance engineering laboratory, and it seemed everything was eventually hooked to that single mission. It made him feel more secure.

"How about some ice cream?" Jerry wondered. The waitress told him he was out of luck, so we walked down the block in the dark to the Hippo, a coffee shop popular with college students. The daytime heat had dissipated, and people were out and about on the street. We ate our cones on a bench outside.

At Los Alamos, where basic research has always had a place, Jerry had been plagued with moral qualms.

"It bothered me a lot more at Los Alamos than it does here," he said. "There I was always consciously trying to relate what I was doing to weapons in order to get money to do it. I think that in order to get money to do research you basically have to prostitute yourself." At Sandia, the managers do that bit of dirty work for him, and he can focus on his research.

"I certainly do have moral reservations about it," Jerry said. "I don't know if I could really work in a design group, that designs weapons, where you're trying to maximize yield or whatever the hell it is you try to maximize. It, to me, seems just a little too close to pulling the trigger. Intellectually, I don't think there is a difference. But somehow morally there is a difference." He looked at the dripping cone. "I find it very hard, with these kind of things, to separate morality from enlightened self-interest."

6

NUCLEAR
DREAMS

.
.
.

If one of the big black crows that soar through the clear blue skies above the desert were to fly two hundred miles south and east of Los Alamos, its keen red eyes would eventually spy an obelisk in the center of a very shallow crater. It is made of black lava rocks tightly cemented into an elongated pyramid and bears a small metal plaque with the inscription "Trinity Site, Where the World's First Nuclear Device Was Exploded on July 16, 1945."[1]

The lava rock comes from an area near the site on White Sands Missile Range. Officially established one week before the Trinity

---

[1]The Trinity Site explosion was only the first of three in New Mexico. The other two were underground explosions. Project Gnome in 1961 was designed to test if the heat from an underground explosion could create steam for use in industrial turbines. The twenty-six-kiloton nuclear device in Project Gasbuggy in 1967 freed natural gas trapped in underground rocks as part of the Ploughshare Program, the AEC's ill-fated attempt to find peaceful uses for nuclear bombs. See Morrissey, "N.M. Has Been Site of Three Nuclear Explosions," *Albuquerque Journal*, March 26, 1986, p. A-6.

Test, the missile range today proudly advertises itself as America's largest military installation. Included within its four thousand square miles is a stretch of desert so arid and forbidding the early Spanish settlers named it Jornada del Muerto—the journey of death.

The black rock monument was erected on the twentieth anniversary of the Trinity Test. Ten years later the National Park Service made it a national historic site. Since the end of World War II there have been calls for a more elaborate public monument to be administered by the Park Service. The proposals are doomed, however. Public parks are supposed to be open to the public, but the Trinity Site lies within the flight path for rocket tests at the White Sands Missile Range.[2]

The Trinity Site has been open to visitors only during annual tours sponsored by the Chamber of Commerce from nearby Alamagordo, tours traditionally held on the first Saturday in October because the intense heat of July could ruin any anniversary ceremony. In 1986, faced with upwards of three thousand visitors a year, officials from the missile range agreed to a second annual tour date, in the equally mild month of April. Alamagordo is actually a good sixty miles from the Trinity Site, but the town has few attractions, and the Chamber of Commerce appeals to the geographically confused with the slogan "Have a Blast in Alamagordo!" Locals joke about setting the tourists straight: the Trinity Test did not take place in a downtown parking lot.

After entering the Stallion Gate at the northern edge of White Sands, it is a seventeen-mile drive to the crater. On July 16, 1985, the stream of incoming cars was efficiently diverted into neat rows for parking by dusty men in military uniform. Despite everyone's worst fears, the temperature never even got into the nineties. About fifteen hundred people went to see where, shortly before five-thirty in the morning forty years earlier, a great blast had lit the sky in an early dawn and shattered windows 120 miles away.

In shorts and sun hats, the visitors walked from the parking area

---

[2]See Ferenc Morton Szasz, *The Day the Sun Rose Twice: The Story of the Trinity Site Nuclear Explosion July 16, 1945* (Albuquerque: University of New Mexico Press, 1984), esp. ch. 9.

to the site down a path of fine gypsum dust that reflected harsh white light from the midday sun. No one said much on the trek toward the little wooden booth sheltering the White Sands Public Affairs staff. Behind it lay the huge damaged hull of poor Jumbo. People fixed their eyes on their feet until they spied Jumbo. "That must be it," they said, before they saw the fence at the end of the path.

That chain link fence went up shortly after the war, when ranchers north of the site discovered strange burns and discolorations on the hides of their cattle. The area inside the fence seemed identical to that outside except that the land inside slopes like a saucer to a depth of about ten feet in the center. The depression from the blast, roughly four hundred yards in diameter, might be mistaken for a natural dip in the crust of the earth by an uninformed visitor or passing crow. The lava rock monument marks the place once occupied by a hundred-foot steel tower, nicknamed "Zero," on which the bomb was mounted. The force of the atomic explosion vaporized the tower. Below it, at ground zero, the bomb compressed the earth as if a giant had pressed its fist onto a wad of clay.

Most of the fortieth-anniversary visitors to the Trinity Site probably did not know that the earth was compressed rather than blown up. The Stallion Gate guards had handed out a ten-page pamphlet on the history of the site. It included a brief statement about the relatively low level of radiation at ground zero and a simple history of the Manhattan Project, the bomb, and the Trinity Site.

But once inside the fence everyone seemed to forget the stapled booklets clutched in their hands. They invariably stopped and looked for something to *look* at. The stone marker was a magnet for their attention, but its short epigram took only a few seconds to read. People posed beside it for pictures. A replica of Fat Man had been hauled in for the occasion. It sat next to ground zero on a small trailer. Kids climbed up beside it and tried to wrap their arms around the dull black surface. Strangers obligingly snapped photographs for family albums.

After looking at the marker and the bomb, most people gravitated toward a low metal and glass structure that looked like an old-fashioned cucumber frame. Inside was dirt, but different dirt than that under our feet. When the Trinity bomb exploded, it fused the desert sand into a glassy green substance, trinitite. Select groups of visitors after the war took pieces of it home as souvenirs, but with time and the elements the glass broke up, and the Atomic Energy Commission eventually bulldozed the site and carted away the remains of the fused sand. The cucumber frame was supposed to preserve a small area in its original state.

Tourists craned their necks and squinted, circling the glass structure slowly like mourners at a state funeral. The trinitite beneath the glass had also disintegrated and was half-embedded in fine desert sand. The most sharp-eyed of the circling visitors pointed out the occasional bit of gray-green stuff, worn dull like sea-glass.

The other attraction at the fortieth anniversary of Trinity was a series of posters wired to the inside of the chain link fence. They briefly summarized the important facts about the Trinity Test. The posters had big pictures and scant text. They were easily read at a slow walk along a short arc of the perimeter. After that, people were on their own.

No one seemed to know what to *do* at the Trinity Site. Eventually some would take a military bus two miles down the road to the McDonald farmhouse, where the bomb had been assembled by a team of Los Alamos scientists. There a middle-aged Hispanic woman told the stranger beside her that her father had heard the blast and seen the light. He'd been so frightened, she said, that he ran inside and shook his children awake. People lingered nearby, hoping to hear more, but she had nothing to add and was quickly carried away by the crowd that pressed through the small rooms of the abandoned farmhouse.

If other people told such stories at the Trinity Site itself they did so quietly. A few prayed openly. Most just wandered around the center of the Site, crisscrossing aimlessly in front of the movie cameras set up by film crews flown in from around the world to record the occasion. A white-haired man wearing a bolo tie with a huge turquoise clasp planted himself next to the obelisk and announced loudly that the bombings of Hiroshima and Nagasaki had saved a million lives. People surrounded him, hungry for information and ideas. Most drifted off when he began a lecture on the safety and promise of nuclear energy.

A younger man from Montreal walked around the site with a geiger counter that clicked like crickets on a summer night. With a video camera mounted on his shoulder, he captured the surprise of curious bystanders as they remembered the radioactivity of the site and realized their reactions were being captured on tape. He said he had no idea what use he'd make of the film, and, grinning, explained that he'd brought his radiation detector "just to be on the safe side."

The Trinity Site transmits history in a purely abstract way. With so little to look at, people helplessly strained to remember details about the war, the bomb, the structure of the atom. The Trinity Site itself was innocuous. The fence and obelisk could easily be moved without most people suspecting a thing. The chattering geiger counter carried by the man from Canada made the abstraction concrete. The obelisk and posters and cucumber frame were obviously mere public relations. The invisible truth was revealed in those arrhythmic clicks. On July 17, when the tourists had gone home and the posters and Fat Man replica had been hauled back into storage, the shallow saucer in the desert would still be ten times more radioactive than background normal.

After people had their fill they walked back down the white path toward the parking area. Those with children waited while their kids, delighted by the echo, ran shrieking through the remains of Jumbo.

•　　•　　•

Scientists and engineers in the weapons laboratories see nuclear weapons as technical devices. Bombs have parts; parts play a role in a process. But everyone acknowledges that the whole, in this case, is truly greater than the sum of its parts. So how to understand the whole? Some turn to poetry.

"These are nuclear dreams," Ed Grothus declared. He liked the sound of that and repeated it for effect: "Nuclear dreams."

For as long as anyone can remember, Grothus has played the

loyal opposition in Los Alamos. Now in his early sixties, a solidly built man with a ruddy complexion, a full head of kinky gray hair, and an air of constant agitation, Grothus went to work as a machinist at Los Alamos Laboratory shortly after the war and stuck with it for twenty years. "Got rich enough to retire," he claimed, but Grothus never retired and probably never will. He became an entrepreneur instead, building an empire in Los Alamos. I first met him in the Shalako Shop, a big gift store filled with Indian trinkets for tourists and better items for the more discriminating buyer. His cluttered desk and a bank of ancient filing cabinets were in a corner of the shop, elevated on a wooden stage so he could keep his eye on things.

Grothus also owns the Los Alamos Sales Company, a complex of five warehouses scattered throughout town. He filled them with used office furniture and scientific equipment he bought at bargain prices from the Los Alamos salvage yard where Gene had found his ruby. The salvage business may have gotten out of hand. Grothus has nicknamed his warehouses the "black hole." Things fall in and never come out. This is no way to run a business. Nonetheless, Grothus proclaimed himself a millionaire. It was not clear where the money was—in inventory? Buried in coffee cans in his backyard? Either seemed possible.

"I was morally outraged by the Vietnam War—free-fire zones, carpet bombing, search and destroy, exfoliation," Grothus said. He started protesting the war early. "'64, '65, '66, '69, I'm raising hell, we've got a group called Los Alamos Citizens for Peace in Vietnam, about thirty-five people," he remembered. (There has never since been so large a peace group in Los Alamos.) Grothus's loyalty was called into question; people wanted him fired, and the laboratory put him through a special security check. Even worse, from his point of view, was the complicity of the majority. "Why weren't the moral leaders of the community as outraged as I was?" he wondered, and stopped going to church. He has been angry ever since.

Grothus summarized the reason for his anger and fear in one word: "cliocide." It is not in the dictionary. Grothus obligingly defines it as "the killing of history." *That* is why he has so loudly denounced the arms race to the people who help design it. But he assigned laboratory scientists no special culpability, and was surprisingly tender in defending them. The scientists at the weapons laboratories are no worse than everyone else, he explained. Because everyone is threatened by the arms race, everyone is responsible. Thus everyone must do their best to stop it. To that end, Grothus has made himself a nuisance in Los Alamos.

Sandia National Laboratories has its own nuisance. His name is Chuck Hosking. In 1983 he began a daily vigil outside the gates of Kirtland Air Force Base, with banners expressing his opposition to the weapons laboratory. Hosking believed that Sandia employees were essentially in it for the money. Ed Grothus doubted that explanation. "I have watched these people so long, I've watched them with their nuclear dreams," he said, savoring the phrase. He too felt the magic of science, the beauty and precision and power of it. His overstuffed warehouses testify to his own appetite for scientific gadgets. Grothus understands the spell.

Hosking has tried appealing to Sandia workers with moral aphorisms such as "Resist Enemies Nonviolently, Like Jesus, Gandhi, and King." Grothus has developed a similar strategy, producing even more charged imagery. Grothus does not use tough-minded technical or political arguments to make his points, even though he must have once used such skills as a technician and businessman. In dealing with moral issues, and particularly the arms race, Grothus is a fierce warrior poet.

Grothus has tried to incite people to share his moral outrage by publishing letters to the editor and opinionated advertisements in the *Los Alamos Monitor*, the inadequate and bland local newspaper. Show the slightest sympathy and he will send you copies of his bitter and angry poems. These decry deceptive politicians, hypocritical churches, and the most profound innovation that affects the human condition, the threat of nuclear devastation. Grothus has gone to meetings and shown up at demonstrations outside the laboratory. He has relentlessly revealed every idea and insight to the broadest possible audience. One year Grothus strung a banner outside the Shalako Shop: "We Are the New Buchenwald." Later he went to the Los Alamos City/County Council with a proposal to change the name of the town to Nuevo Buchenwald. It may be the most unforgivable in the string of offenses he has committed against Los Alamos.

His battle-weary neighbors now politely ignore his messages. Most resent his efforts to be, as one scientist put it, "the conscience of the community." They see him as well intentioned, dedicated, and loony, a latter-day village idiot who proves the tolerance of the community and might even be a charm against others of his kind. Grothus explained their reactions differently, though. He thought them captives of their nuclear dreams.

If science and technology are rational, cool, and objective, then what fuels these dreams? Why would Galileo risk his mortal soul for the sake of a theory? Why did the faces of scientists from the weapons laboratories invariably light up when they talked about their research? If scientists struggle to achieve objectivity, why did Karl yearn to feel the heat?

Scientific work requires more than just putting together facts and working out the logical implications of ideas. What makes research exhilarating to scientists is the experience of discovering something, ordering things in a new way, seeing things as no one has ever seen them before. "There's a real aesthetic to science," one Los Alamos physicist said. "You have to appreciate it on its own terms." Then, as if he could not stand the seriousness of it, he elaborated, "There's a certain charm about exploring the universe." What is that charm? He mentioned Newton and Einstein. What he meant was "discovery."

Einstein never did much, scientifically speaking, after he developed his General and Special Theories of Relativity as a young man. Yet his drooping features, haloed by frizzled white hair, mean "science" to Westerners as surely as a cross signifies Christianity. Everyone knows that Einstein renamed space and time. His ideas were counterintuitive, but they were right. He was inspired, and his inspiration changed everything.

Of all the mysteries of consciousness, creativity is the most intractable. Having an insight sheds light on something, but the source of the creative experience lies in shadow. It seems that people just "get" ideas. The new thought comes as a shock. It dawns on us, or we are struck by it. It intrudes like a stranger at a family gathering. So powerful is this sense of intrusion that some people are convinced that the only possible source of creative insight is God; nothing so remarkable could plausibly come from our mere selves.

Only two social roles in the Western world expect and applaud insight, inspiration, and creativity. Only scientists and artists of various stripes are supposed to be consistently clever in this peculiar way. We "know" that painters must paint and poets must struggle to convey their inner vision. When they succeed in moving us they are heroes, albeit often tragic. When they fail, they are pitiful. In popular imagination all artists are crazy, or nearly so, and represent the barely constrained creative force, conceived of as a kind of wildness.

Scientists, on the other hand, are thought to be tamed by their training and constrained by their objectivity. And they *are* disciplined, learning to talk as if only logic and the facts before them really matter. But the creative moment is what gives science its power. The drudgery of laboratory notebooks forgotten, scientists celebrate with the joyful shout "Eureka!"

These are not mundane discoveries. They may concern natural and earthly truths, but nature has a supernatural beauty to the scientists at the weapons laboratories. John Manley saw Haley's Comet when he was a tiny child. The schoolteacher in his hometown of Harvard, Illinois, was a boarder in the Manley home. She took little John outdoors one night, he remembered, held him up, and pointed out a long white streak in the sky. It might have been the moment the universe impressed itself on him. A younger Sandia scientist gave his mother credit for his own reverence for nature. "She would point and tell me, 'That's the moon,' and tell me a little nursery story about the moon," he recalled. "Before I even started school, she had taught me the constellations. Of all the things that inspire awe in my soul, it's the night sky. Even as a child, I knew that they were suns, larger than the sun that came up every morning." A new discovery or understanding of nature can change the way you feel about everything, they explained.

Poets and artists strive for the same effect. They want the world to seem different after the poem or performance. People who visited the Trinity Site on its fortieth anniversary knew it was only an open field on a missile range, but they went anyway, to see a place where the world was changed. People who go out of their way to visit Los Alamos have the same motive. Something important happened there. A visit might make it all clearer.

By now there must be millions of poems about the nuclear age. Like resentful diatribes against the Grim Reaper, most are probably embarrassingly impassioned. Grothus could not understand why his metaphors fail to inspire the same frantic terror in his Los Alamos neighbors that continually energizes him. He figured they must be mesmerized by a more appealing version of the same strange dream.

At Sandia another poet, Tom Grissom, also despaired of communicating with his fellow scientists in anything other than technical terms. When I visited him he was polishing the preface for his second poetry book. We chatted about writing: How do you choose the right word? How do you evoke emotion? Grissom, a polite southerner, soon interrupted my ramblings. He was so *relieved*, he said, to talk to someone who appreciated the mechanics and aesthetics of writing. Most people at Sandia would find our conversation nonsensical. They understood science and engineering, but hadn't a clue about poetry. It was one of the things that had convinced him to quit his job. His colleagues at the weapons laboratory may love studying the details of nature, but they seemed deaf to the sweetest sounds.[3]

• • •

Walt had broken his arm months earlier in a motorcycle accident. With his right arm in a cast, the forty-four-year-old mechanical engineer could not drive his car, so I picked him up outside his

---

[3] Both of Thomas Grissom's books contain poems reflecting on his experiences at the weapons laboratory. See *Other Truths* (Francistown, NH: Golden Quill Press, 1984), and *One Spring More* (Francistown, NH: Golden Quill Press, 1986).

Los Alamos National Laboratory office for our luncheon interview. The talkative redhead groused good-naturedly about his handicap as he got into the car and then led me on a roundabout tour of Los Alamos to be sure I understood what people meant when they said they loved the mountains and pines. It was early May and the view of the orange canyons was filtered by the light green leaves of trees growing along the top.

"The ocean here is the air," Walt said, throwing out the first of many facts and opinions on a bewildering array of topics. The man characterized by one of his office mates as "the horniest guy on the Hill" manfully tried to censor his language. He grimaced with embarrassment at every failed attempt to convert "fuck" into "fudge."

Walt had made his career as a "contract engineer," working for firms that offer expertise on short-term defense research projects. The short term can stretch itself out, however, and since the late 1970s Walt had lived in Los Alamos, helping the lab's regular staff members design and draft parts for their very large lasers. His work had implications for Star Wars, Walt said vaguely. He approved of Star Wars. It was consistent with standard military wisdom: Take the high ground.

Walt loved working at Los Alamos. "Easiest job I ever had," he explained. In a typical eight-hour day at the laboratory, Walt figured he spent three hours actually working. The other five he slyly labeled "thinking," which turned out to be coffee breaks, chatting with the guy at the next drafting table, planning and playing with ideas. Nonetheless, Walt was starved for conversation and found it "really neat to be able to talk about this." In retrospect, looking at my notes, I saw that "this" was a seamless web: global ecology; the natural limits on nonrenewable resources like oil; politics; philosophy, physics; morality; the media; his family; Einstein; the nature of reality.

Walt's right arm was immobilized, but he was left-handed, and two weeks after our interview sent a five-page handwritten letter elaborating his thoughts about the social responsibility of members of the scientific community.

"Those of us who began our education just after the end of WWII were well aware of the great promise of the 'postwar era' and the potential for good of technical developments in all areas of civilian life," he wrote.

"Unfortunately, as an army brat of a career officer (my dad served duty in three wars—WWII, Korea, and Vietnam) I gained *firsthand knowledge* of what reality was going to be like, and there wasn't much anyone was going to be able to do about it."

The problem, Walt explained in spiky uppercase letters, was the "Malthusian starvation limit."

"Too many peoples vying for too few resources tend to precipitate conflict. At age ten, spring 1950, Clark Field Air Force Base, Philippine Islands, I sat on our front porch and watched P51 Philippine Air Force fighters bomb and strafe Huks in the hills surrounding the base. Summer of 1950 my older brother and I would hang on the fence out by the model hobby center and watch squadron after squadron of brand-spanking-new F-86 Sabre jets head out for combat in Korea."

His brother immediately decided to make a career of the military. "I wore glasses and knew it wasn't in the cards," Walt wrote. "However, I thoroughly enjoyed building model airplanes and discovered science fiction that summer of 1950."

Walt adored the fantastic devices, the gee-whiz weaponry, and the expanded and distanced perspective on the earth and its inhabitants. Stardust permanently altered his vision.

Sitting over lunch, Walt confessed, "I formulated all of my ideas, basically, before I got out of high school." He still felt a thrill, he said, when he looked at the pictures of the first astronauts on the moon taking pictures of the earth. Robert Heinlein's robots and space warriors occupied his adolescent dreams. I was startled. I told him that I, too, had been a science fiction fan, but that I was drawn to the aliens. Walt looked confused, laughed uneasily, and continued his story.

The romance of science fiction was only part of his decision to be an engineer. "I figured that if there's another war, they ain't going to be bringing the engineers and scientists home in body bags," he drawled. "I saw men I knew coming home in body bags when I was a kid. I was thinking, when I was in junior high, if there's another war they ain't gonna send the guys that can design the weapons out on the front line." He wanted to help conquer the universe, but he did not want to be a casualty of the quest. "So," Walt concluded, "I decided I'd be a weapons builder, not a weapons user."

But the weapons laboratories have their fair share of weapons users, too. One man at Los Alamos never had a doubt about the morality of working in a weapons laboratory. He had been stationed in Europe during World War II. "I had only three months in combat but that experience influences a lot of my thinking," he explained. "There was one day when our unit went from 167 of us to 40. I was so stinking happy that I didn't have to go to Japan."

Another had a similar experience and the same reaction. "Any comments I have about the morality of weapons is informed by the fact that I was sitting in the Philippine Islands waiting to be part of the invasion of Japan," he said. "And there was considerable joy in the Philippine Islands when the bomb was dropped on Japan, I can tell you." Strategists had predicted a million casualties from the impending invasion, a third laboratory worker reminded me. "I was in an amphibious group, and our job was to land on a little island near Japan," he recalled. "If it hadn't been for the bomb, I know where I'd be yet."[4]

Walt remembered men returning to the base in body bags, but the death burned into his brain forever was more like a dream than a real event. "I saw this fish story with my own eyes," he said, leaning back and resting his cast on the back of his chair. "A young whale got trapped in a lagoon in the Philippines. This was the Korean War, and there were army guys blasting away at the baby whale. They thought it was a black submarine." Walt laughed. "I'll never forget watching these fifty-caliber tracers ricocheting across the bay, and this long stream of red blood across the lagoon."

He ran the fingers of his good hand through his hair and shook his head. "They killed the baby whale. They hoisted it out of the water. But then the crane broke, and as it came down it killed three Filipinos."

---

[4] See also Paul Fussel, "Thank God for the Atom Bomb. Hiroshima: A Soldier's View," *New Republic*, August 22 and 29, 1981, p. 28.

The blood and bullets and science-fiction stories all pointed in one direction. "I realized when I was ten years old, your only social responsibility is to keep yourself alive and avoid pitfalls." Walt looked at me sideways and added, "Um, and try to contribute. Probably the majority of people in the world, they want a good job, a good woman—I'm talking about guys, not gals—and a good car.

"I see myself as a voyager in time and space," Walt continued. "I just happen to be in this century, and I'm just trying to get through it. I really have no control over it," he said.

• • •

Walt saw the future early. Others at the weapons laboratories were surprised by how quickly their ideas changed.

"When I was finishing up my postdoctoral research, my phone lit up and the government labs were calling me to come out for an interview. I got phone calls from people I never heard of. They were all calling within three weeks. I was shocked. I didn't think I'd get a job."

Arnold was explaining how he had ended up at Sandia Labs in the 1980s. In his early thirties, he was slightly overweight, slightly stooped, his blond hair already thinning. The secretary for the group he works in had called to set up an interview. I thought he must be some elderly eminence, but no—the secretary was a motherly sort who felt that as the new kid in town Arnold needed to meet more people, "especially more women," he laughed apologetically, "especially women from outside the lab."

He had bought a big house not far from the base, close to the shopping malls, and furnished it with an early American orange-and-brown print couch and matching chairs.

He'd grown up in the rural Midwest, he told me, sinking into a tan Naugahyde recliner from which he would not budge for four hours. He was the son of a small farmer, but ten years of college and postdoctoral research had taken him all over the country. In graduate school he'd had vague thoughts about getting a teaching position at a university somewhere, but he was pessimistic about his career prospects because his specialization in space physics is relatively rare. So, when his phone started ringing off the hook with offers from the national laboratories, Arnold saw dollar signs.

Arnold was a bit modest about the money. His starting salary approached $47,000, which was nearly double a typical income for an assistant professor at a state university. But it really wasn't the money going into his own pocket that had lured him to the lab, he said.

"To pursue my research in a university environment I'd have to bring in an incredible amount of money through grants and contracts," Arnold explained. "To pay for my computer bill I need, probably, $100,000 a year, which is a lifelong commitment to writing proposals." Arnold hated writing proposals—"too fragile." All that time and energy explaining to the National Science Foundation why it should give you a big grant might be wasted. The project might not work right, might not get good enough results to be funded again, and then you're out in the cold with a $100,000 computer bill. If you love to do research, Arnold said, the labs are nearly perfect.

"I like working on problems that are more than truth and beauty," he added. "I like working on problems that have an impact on society and its future."

Firmly ensconced in his Naugahyde easy chair, Arnold considered the impact of his work. His political views had shifted 180 degrees since his undergraduate days. "I never would have worked on anything that was defense related," he said. "I had a lot of justifications for it at the time, but the real reason was that I didn't like the Vietnam War and I didn't want to go fight."

Arnold was saved by a student deferment. Later, he began reading *Time* magazine for a broader perspective on the world. He concluded that both individual and national self-interest led to aggression and to the classic choice: fight or surrender. "And I don't believe in capitulation," Arnold said firmly.

The Soviet people just want to raise their families, drink vodka, and have fun, Arnold thought. But the Soviet government seemed truly evil. "I think they really have malicious intent," Arnold said.

"I believe in truth, justice, and the American way." Sandia helps protect the only bright star in the political world.

Arnold knew exactly what he did not like about working at Sandia, too. Paradoxically, the *important* projects were his pet peeve. Important projects command a lot of manpower. "I become a small cog in the machine," Arnold complained. "I prefer to stay on the smaller, harder projects. It keeps you awake at night but it's exciting."

"It keeps you awake at night" is a fairly common figure of speech. Arnold used it repeatedly, and I thought nothing of it. But just before midnight, as I prepared to leave, I learned that Arnold meant it literally. It was a weeknight, and in a few short hours his alarm clock would ring. I apologized for losing track of the time. Standing in the darkened doorway of his house, Arnold shrugged and admitted to insomnia. He stayed up because his mind was filled with thoughts of his work.

"For the first time in my life," Arnold said, "I worry about whether my answers are right or wrong. Not because someone's going to tease me about it, but because it could have very serious consequences if I'm wrong."

The number of people working on a classified research project is so small, Arnold explained, that everyone might make the same mistake. If the mistake was serious, an important weapons project might not work as planned. The farm boy from the Midwest who did Star Wars research felt the burden directly, as stress and muscle pains in his chest and shoulders.

"It really worries me," Arnold said softly. He looked weary. "I lose sleep over this frequently. Aside from losing sleep, I feel I'm better off now than I was before. And really I'm pretty happy with the whole thing."

7

THE
BOGEYMAN

•
•
•

The Land of Enchantment has few of the attractions usually sought by tourists. There is some good skiing in the northern mountains, but Colorado has better. There is no ocean—except for the air—and the public beaches along the edges of the artificial lakes and reservoirs are unpleasant. The vistas are breathtaking, but only if you happen to drive by when the sun lies near the horizon. At midday the brilliant oranges and reds fade to brown and tan.

From the interstate highways the desert looks like an empty unshaded parking lot outside Dante's Inferno. Only after you get out of your car and slither between the strands of barbed wire lining the highway does the pattern of desert life reveal itself.

Close up, what looks like a small grayish wad of paper turns out to be the undigested remains of a mouse, tiny bits of bone embedded in matted hair. The stem of gray-green plant has a shining black bump on it. The red hourglass on her belly betrays the black widow spider, a small spot of poisonous beauty waiting at the edge of her messy web. A scattering of gray sticks becomes a pile of bones splintered by a carnivore's teeth. A rusted piece of tin once held a cowboy's tobacco. The brown streak beneath it is a hawk's striped tail feather. A flat gray stone is really a shard of Indian pottery. A round white rock has empty eye sockets.

These details are easily missed. Tourists generally seek the obvious. In New Mexico, that means "multicultural diversity."

Indians may fit a single stereotype in the Wild West movies, but in New Mexico it becomes apparent that the pueblo peoples, the Apaches, the Navajo, and the Hopi, are all quite different. The natives of the Southwest have held onto many of their traditions despite staggering poverty and a tortured history best summarized as efforts by the federal government to kill their cultures.

"Everyone said in the forties and fifties that the Indians would be culturally assimilated," said Chris Dietz, an anthropologist in the northern part of the state. Dietz has studied the impact of Los Alamos on the indigenous people and found that, contrary to most anthropologists' predictions, the Indian cultures have persisted and become stronger. "Not only that—they have evolved and elaborated new variations on their traditions, new dances and so forth."

The Indians point proudly to their successful rescue efforts, but Dietz believed that two pueblos—Santa Clara and San Ildefonso—owe the revival of their traditional lifestyles to Los Alamos National Laboratory. Both are the darlings of collectors, producing beautifully shaped and incised black pottery. Twenty years ago both small communities were nearly dead. Young Indians, encouraged to assimilate into the mainstream culture, grew up and moved to the cities. When the lab started hiring Native Americans as service workers and technicians in the 1970s, people from the two nearby pueblos could afford to stay in their traditional homes.

The Hispanic community includes people who proudly trace their lineage to Spain and, in officially bilingual New Mexico, speak a style of Spanish that linguists have traced to their sixteenth-century conquistador ancestors. Many like to consider themselves pure Spaniards and are insulted to be called Mexican-Americans, thinking it denigrates their forebears—Indians say because it hints at the inevitable quantum of Native American blood. Others, with equal pride, call themselves Chicanos, celebrate Cinco de Mayo, a Mexican national holiday, and seek a unified Hispanic political movement. Young Hispanic low riders equip lovingly restored classic cars with hydraulic suspensions and glide bumper to bumper down city streets in a mechanized version of the Spanish promenade. An unknown number of Mexican nationals work with (and sometimes without) the green card that legalizes their presence.

In the Land of Enchantment, if you're not an Indian or an Hispanic you must be an Anglo. "Anglos" may be cowboy ranchers, retirees from "back East" (anywhere east of Texas, that is), hippies in solar tepees, mystics who have found the navel of the earth, white-turbaned American Sikhs headquartered just down the hill from Los Alamos in Española, or weapons scientists. The promise that tourists will be fascinated by the state's multicultural diversity is no lie.

Unfortunately for the tourists the most picturesque traditional Hispanic and Indian communities are also very poor. "I get very annoyed when people talk about the 'Land of Enchantment,'" Dietz insisted. "The reason why the land here is so enchanted is because there are no jobs here. The enchantment produces poverty, illness, misery, and despair. Look at Navajo country. The kids up there are shooting themselves, and the Navajo elders say one of the biggest problems they have is Satanism among their kids. Look at the suicide rate in the state." The rate is consistently about twice the national average, and highest in the poorer counties.

The disparity in wealth is obvious and startling. Tourists in Santa Fe are charmed by the manicured Spanish-style plaza, the soft flow of adobe forms, and the lovely shops and crafts galleries filled with very expensive goods and impeccably-groomed sales clerks. But visitors who take refuge from the sun in the cool hotel bars get nervous when approached by the white-haired Indian men in traditional dress who are trying to sell the waterfalls of silver and turquoise necklaces displayed over their forearms.

An economist at the University of New Mexico once likened the state to a Third World country. Most of the people are peasants. Their traditional lifestyle is hand to mouth. The economy in urban areas depends on dollars dispensed by the U.S. federal government.

The federal money in New Mexico even fits into the traditional categories. Some is for social welfare and economic development projects. For example, the Navajo, the largest Indian tribe in North America, get about half of their annual budget from the federal government. One in five residents of the state lives below the poverty line. Excluding the richest three counties (Los Alamos, Santa Fe, and Bernalillo, where Albuquerque is located), an average of 30 percent of New Mexico's people live in poverty.[1]

Much of the federal money comes in the form of military aid. In 1986, for example, the departments of Defense and Energy jointly pumped $3.26 billion into the state's economy, helping to make New Mexico among the top four recipients of federal funds per capita. Those dollars flow to military bases and DOE installations throughout the state and up and down the Rio Grande Valley, designated a "high technology corridor" by ambitious state planners. The corridor runs from Los Alamos in the north, through Albuquerque, to the White Sands Missile Range in the south.[2]

Between the desert and the multicultural diversity, even the insatiably curious have enough material for a lifetime of study. With so many cultures to explore, few tourists concern themselves with the weapons laboratories. Residents of the state generally watch the laboratories in the same way casual investors monitor the Dow Jones industrial average. Newspaper headlines present the gist: Is some new debate Good for the labs or Bad for the labs?

"Good" for the labs, not surprisingly, means more federal money, always welcome in a state consistently ranked among the poorest five in the nation in per-capita income. New Mexico has received more money for SDI research per-capita than any other state, for example. When the press reports details about the laboratories' work, they generally stick to stories about scientific breakthroughs in "basic research"—the kind of work that is not "mission

---

[1] Figures based on analysis by the New Mexico Conference of Churches, "1 in 5 in State Live in Poverty," *Albuquerque Tribune*, June 12, 1986, p. A-2.

[2] Data from Morrissey, "Defense Dollars Drive New Mexico Economy," *Albuquerque Journal*, May 10, 1987, p. A-l.

oriented"; in other words, not weapons work. The bomb research generally gets headlines only when dollar signs are at stake.

So, for example, in 1988 and 1989 the Department of Energy was rocked by scandal. Its weapons manufacturing facilities around the country were "wearing out," headlines screamed, and many were polluted with radioactive and toxic wastes. Groundwater at Sandia and Los Alamos had been contaminated with radioactive material, at levels exceeding the recommended amounts for drinking water. The DOE estimated it would cost between $2.8 billion and $3.3 billion to clean up as many as three hundred sites at Los Alamos, a project that could take as long as twenty-six years. Cleaning up toxic and radioactive wastes at Sandia would take about twenty years, at a total cost of $486 million to $531 million. And the Inhalation Toxicology Research Lab on Kirtland Base— the Puppy Palace—would also need nearly $50 million, at minimum, to deal with its toxic wastes.[3]

Within a few months the DOE had confessed that mismanagement dating back to the 1940s had led to extensive environmental contamination and chronic safety violations, most of which were previously hidden. Critics charged that the DOE's timely admissions were calculated to convince Congress to increase its budget. Indeed, by mid-year New Mexico's Senator Pete Domenici, the ranking Republican on the powerful Senate Budget Committee, made headlines by promising that the DOE's woes would benefit Sandia and Los Alamos. The two labs were expected to help provide technological solutions to the DOE's toxic and radioactive trash problems nationwide.[4]

• • •

In 1985 the Strategic Defense Initiative was all the rage, and President Reagan and Soviet leader Gorbachev were heading to

---

[3]U.S. General Accounting Office, GAO/RCED-88-197BR, *Dealing with Problems in the Nuclear Defense Complex Expected to Cost Over $100 Billion*, July 1988; GAO/RCED-88-229FS, *Supplementary Information on Problems at DOE's Inactive Waste Sites*, Sept. 1988.

[4]Keith Schneider, "DOE's Candor Is from Fear of N-Plant Mishap," *New York Times*, reprinted in *Albuquerque Journal*, Sept. 14, 1988, p. A-7.

Geneva to discuss arms limitations. In late September a coalition of peace groups in Albuquerque held a press conference to criticize the former and applaud the latter. One of the speakers was Bill Pletsch, a Ph.D. student in mathematics at the University of New Mexico and the leader of the school's tiny chapter of Educators for Social Responsibility.[5]

Pletsch began his statement to the press by announcing a rumor that had been floating around the campus: "Sandia nearly lost Albuquerque!"

"Someone at Sandia Labs was going to do some testing on an unarmed nuclear warhead," Pletsch charged. "As it was, the warhead was not unarmed." Pletsch thought people in Albuquerque ought to be worried about what went on behind the fence surrounding Sandia.

Pletsch is a bookish man in his late thirties, the kind of person who can think about nuclear war without panicking but who becomes acutely agitated when he can't get a napkin out of the dispenser at the university cafeteria. Later Pletsch confided that he'd publicized the rumor of near disaster to see how the press would react. He had allowed himself to be talked into it against his better judgment. Not that he doubted the rumor, he added quickly. He thought it was probably true.

How could such a thing happen? Sandia workers, in conjunction with scientists and engineers from the Los Alamos and Lawrence Livermore national laboratories and members of the military, sometimes run tests of the non-nuclear components of nuclear weapons. These tests verify the operation of all parts of the process that detonates a weapon, up to, but not including, the fission and fusion explosions in the core of a thermonuclear bomb.

To conduct such tests of the "peripheral parts and components" of a bomb, the materials that are supposed to undergo fission and fusion are removed and replaced by a "Joint Test Assembly," or JTA. The JTA is a package of measuring instruments inserted into

---

[5]I was faculty sponsor for that now defunct organization. It never had enough members to be officially recognized by the university's student government association.

the space normally filled by nuclear explosives. (The JTA is so named because the administration of the testing program is jointly supervised by Sandia and the U.S. Air Force.) In the Mathematics and Physics departments at the university, people were saying that a Sandia technician, who happened to be in the vicinity on an unrelated mission, fortuitously noticed that the weapon about to be tested had not been disarmed.

The press's response to Pletsch's report was a telling indication of how much the community trusts the laboratory to act responsibly and with unerring competence. Only the university's student newspaper, the *Daily Lobo*, covered the press conference and Pletsch's statement. The Public Affairs Office at Sandia said that only one newspaper (but not the *Lobo*) had even asked for its comment on the rumor.

Pletsch feared mortal embarrassment for making so much of an unverifiable story. Since his statement was widely ignored, his reputation in the community at large was safe. This gave Pletsch little peace of mind. For all anyone knows, he pointed out, Sandia nearly blows up Albuquerque every week.

Pletsch's name was mud at Sandia National Laboratories, however. The labs printed a commentary on his rumor in the weekly *Sandia Lab News*, in a column titled "Antojitos," Spanish for "Tidbits," under the headline "Drowning in a Sea of Naiveté?"

> Apparently there are educated people out there who believe that armed nuclear weapons are simply lying around in a bin, like jellybeans, and it's possible for someone to pick up a live one inadvertently! While we obviously don't want to describe the precise means of protecting weapons from unauthorized users, suffice it to say that weapons are under constant surveillance of one kind or another. And no complete nuclear weapon ever enters Sandia Labs.[6]

"Protecting weapons from unauthorized users" is, in Department of Energy slang, "nuclear safeguards and security." Nuclear "safety" means making sure a bomb doesn't go off accidentally because its circuits got drowned in a rainstorm or fried by their prox-

---

[6] *Sandia Lab News*, Oct. 11, 1985, vol. 37, no. 20.

imity to uranium and plutonium. Safety is primarily a technical problem.

Safeguards and security, on the other hand, present both technical and logistical challenges. Safeguards and security researchers develop the procedures and devices that keep the bad guys from getting anywhere near the jellybeans. People who had worked in the safeguards groups at Sandia were extremely tight-lipped about their work, but they were proud of helping to protect the bombs from "the deranged," as one man summarized his enemies.

"We're designing systems that are resistant to terrorist attacks," one Sandia engineer said. His specialty was highway and rail transportation of nuclear warheads. Others in his group worked on protecting transport systems for nuclear waste and reactor fuels.

Researchers in safeguards and security are luckier than those in weapons design groups. They suffer less administrative pressure to deliver their technology on a fixed schedule. "If it looks like we'll need more time," the engineer explained, "they'll push the deadlines back." Still, he warned, the current system of safeguards and security was quite effective. "If you try to break in, something bad will happen to you. We try to develop nonlethal, effective deterrents."

Why nonlethal? I wondered. He suggested I imagine an adventurous country boy drawn to a slow-moving train. And he was proud to report that their techniques for protecting the trains, trucks, and nuclear sites have resulted in "spinoffs," or "technology transfer": the application of weapons-related science to other purposes. Some are used to protect U.S. embassies overseas and important government officials.

A few antinuclear activists in Albuquerque monitor the comings and goings of the specially designed trucks that transport nuclear materials, as an extension of the nationwide effort to protest the "white trains"—the trains, now painted in various colors, that transport bombs from their final assembly plant in Amarillo to various arsenals and storage sites. But these are peaceful protests by people with no interest in hijacking or theft.

The rumor that Sandia nearly lost Albuquerque wasn't about the threat of "unauthorized users," however. The people who would be at fault if Pletsch's story were true were Sandia employees. And the thrust of the *Sandia Lab News* coverage of the rumor turned out to be a warning to lab employees. "[A]ll the naiveté is not 'out there,'" the story suggested pointedly.

> I believe that *we* can be every bit as naive as some of the folks I've criticized above, especially when it comes to the temptation to boost our egos by sharing with a friendly stranger at a restaurant or interested acquaintance at a party the excitement or challenge of the work we're doing at the labs.

Lab employees were reminded to respect the principle of "need to know," even with those who have security clearances. "Need to know" means just that—you must have a good reason to get access to classified information.

After Pletsch accused the labs of a near accident and cover-up, I asked the Public Affairs Office for more information. The "Tidbit" about Pletsch ridiculed but did not refute the rumor. While it asserted that armed weapons are secure from unauthorized persons and never enter the labs in complete form, it begged the question of whether *authorized* persons might have contact with armed warheads assembled *after* their components are brought onto laboratory premises. The Public Affairs Director for Sandia responded to my inquiry with a note saying, "Parts are not assembled into completed weapons at Sandia. No one knows where the above rumor originated."

Pletsch's story was treated as a joke exposing the stupidity of those who repeat it, but the fact that it was used as an occasion for reminding lab employees not to talk about their work belies its supposed humor. The rumor was no joke when it leaked. For by extension, whether or not such a catastrophe nearly occurred, no outsider should be made curious enough to investigate.

The *Sandia Lab News* story assumed that it is irrational for outsiders to distrust the lab, therefore discouraging inquiry into potentially unsafe laboratory practices. But at the end of 1988 the press reported that a shipping error had landed part of an artillery shell containing both conventional explosives and depleted uranium in the Sandia National Laboratories site at Livermore, California. Apparently the prototype of the W-82 shell, a battlefield weapon designed to have an eighteen-mile range when fired from an artillery gun, had been lying around the lab for four days in mid-1987 before someone realized it didn't belong there. Sandia officials in Albuquerque explained that the device could not have undergone a fission explosion. And they assured reporters that the odds were against an explosion of the conventional materials.[7]

The *Sandia Lab News* story also hinted that the rumor reported by Pletsch may have originated inside the lab, as Pletsch suspected. After all, why else warn against the friendly stranger and interested acquaintance?

Almost all work at Los Alamos and Sandia requires a security clearance, even though many workers at the two labs rarely get near secret information. At Los Alamos it is easier to do research without a security clearance. Numerous graduate students and foreign scientists go up the hill for summer study projects, postdoctoral research, and consultations. Because the laboratory is spread across nearly thirty thousand acres at the top of the mesa, it has many "open environments" where someone in a nonclassified project can usually gain easy access to the necessary facilities without a clearance. But for uncleared scientists, Sandia, with the bulk of its facilities concentrated inside the security perimeter, is a harder nut to crack.

As they enter and leave the labs, employees show the guards their plastic coated badges bearing their name and photograph. At Sandia, staff member badges have a blue background. Letter codes announce the type of information the wearer may see, with a "W" signifying that the badge holder has access to general weapons information. People who transport tools and equipment from one technical area to another wear an additional green badge, also coded with information about the types of materials they may move and from where to where. What gets moved where in the weapons complex is a serious business. The W-82 shell mistakenly shipped to Sandia Livermore should have been marked as a clas-

---

[7]"Shell Accidentally Sent to Livermore Lab," (AP) *Albuquerque Journal*, Dec. 31, 1988, p. B-3; orig. reported in the *San Jose Mercury News*.

sified package, with its radioactive and explosive contents noted on the outside of the wooden crate.

Laboratory employees are civilians, and their clearances are issued by the Department of Energy. Background investigations are conducted by the FBI and the U.S. Office of Personnel Management. Almost a quarter of the DOE's 220,000 security clearances are issued by the Albuquerque office of the DOE.

The DOE classification system distinguishes among three types of secrets: confidential restricted information, secret restricted information, and top-secret restricted information. The system has five clearance levels (secret, top secret, "L," "Q" nonsensitive, and "Q" sensitive). Access to data that is confidential or secret requires a Q-clearance. All regular laboratory staff members and technicians must get a Q-clearance, usually issued within six months of their being hired. Even so, if someone wants to see top-secret information, he or she is supposed to get special permission from the clearance officer by establishing a "need to know."

Arnold found the secrecy at Sandia annoying. "If I'm working on something that's highly classified, it's very inconvenient," he said. "I can't leave my desk. If I go to the bathroom I have to carry a bundle of papers under my arm."

His complaint was echoed by scientists at both laboratories. "It's a pain in the ass," one Sandia woman said. For those with no experience dealing with classified materials, what counts as a security infraction may come as a surprise.

If you close a safe and leave the dial on the last number of the combination, that's an infraction. If the am/fm radio in your office has a tape recorder in it, that's a second infraction. And if you do classified work on a computer terminal or desktop microcomputer that is not shielded, so that the electronic impulses it emits could be picked up by sensitive electronic equipment and, in theory, be decoded by enemy agents (at Los Alamos they call these "compromising emanations"), that's a third violation. At Sandia, three security infractions and you're out.

Scientists at the weapons laboratories see secrecy as a practical hassle. Sometimes it causes more serious problems, though. Dwayne was a nuclear engineer who got hooked on high technology in the military. In 1961, when he joined the Air Force, he was proud to be one of America's knights in shining armor. As a helicopter pilot in Vietnam, Dwayne was astounded by the new laser-guided smart bombs. They struck their targets with uncanny accuracy, and convinced him that good science made for good warfare.

Public opinion about the military has shifted since the Vietnam War, but Dwayne had kept himself as physically fit as an Air Force cadet and was proud as ever to serve his country. The big problem with national defense research, he said in a soft southern accent, is the secrecy. "Some things need to be classified," Dwayne acknowledged, "but it is possible to pervert the classification system fairly easily and cover up mistakes." In his experience, research that was not published in the open literature was never as rigorously defensible or well documented.

"Most classified projects I know of are very goal oriented," Dwayne said by way of explanation. "As a result, when they get something that works, they go with it." Generating classified documents is a lot of extra trouble, and most people avoid it. Thus what is accepted as common knowledge about certain classes of secret technical information often boils down to engineering rules of thumb conveyed by word of mouth.

"The fundamental problem with that is, you get old wives' tales, that '*this* technique didn't work,' and 'so-and-so showed *that*'— and you can never trace *that* down." Dwayne pursed his lips and then grinned. "In the Air Force, safety was the bogeyman—'If you don't do this, the safety guys are gonna get you.' At Sandia, security is the bogeyman, and 'If you don't do this, security is gonna get you.' "

It's unsettling to think that secret research might be less rigorous than open research. Many Los Alamos and Sandia scientists said they had found themselves cut off from the mainstream of scientific research once they began working on classified projects. One Los Alamos scientist remembered his Ph.D. adviser warning him that he would be lost forever from civilization in the secret society on the Hill. Others told of academic scientists who refused to send preprints of their articles (an important means of communication in the scientific community) to people doing classified research.

Most Los Alamos and Sandia scientists said they were sorry that their secret research did nothing to advance their disciplines. "Scientific progress for three hundred years depended on the free and across-national-boundaries exchange of scientific materials and insights," said one physicist whose work was mostly classified. "Classification tries to close that off, to put boundaries on everything and keep things boxed off from each other. It's just an abomination from a scientific standpoint."

A disgruntled young Los Alamos weapons designer felt that his own group in X-division was so boxed off it had become stagnant. "There's no peer review of the work that's done here," he complained. "There's no jury. People can do very bad technical work here and nobody knows." He thought Los Alamos weapons designers mistook 1955-style physics for state-of-the-art because they had so little contact with the rest of the scientific community. "They're in a bubble, scientifically," he said, adding that he hoped to get out.

Younger scientists at the two labs quickly caught on to the nature of the bubble. It may keep uncleared scientists *out*, but it also kept them *in*. Classified projects cannot be included on a public résumé, for example. Eventually the scientific community loses sight of you, so, as one man said, "The longer you stay in weapons work, the harder it is to leave."

People with security clearances cannot discuss their classified work with their families. To accommodate the personal difficulties that causes, both weapons laboratories have instituted "family days." Once every five years, the secrets are locked up and lab employees can bring their loved ones in to see where they work. The fact that there was no *physical proof* that his physicist father worked at all disturbed one twenty-five-year-old man who'd spent his childhood in Los Alamos. "It just seemed really weird to me," he remembered. "I couldn't tell anybody what my dad did"—because he didn't know himself. The family day made the difference. "He showed me this big computer program he was working on," he said. "I never really had a clue what he was doing at the lab until recently."

Arnold believed his Q-clearance affected his love life. "I had a very naive attitude," he said. "I thought that when I graduated with my Ph.D., I'd be able to walk out into the world and capture a girlfriend effortlessly. She'd throw herself at my feet and feed me grapes and so on. I've learned better."

Arnold was well-mannered, pleasant, and easy on the eyes. There was no obvious reason for him to be a loser in the dating game. The pattern was obscure, but eventually Arnold recognized that people formed their first reaction to him based on his job.

"I've had girls say it would bother them that there were things I could not and would not discuss with them," he said. "I would tell them, 'Look, the stuff I can't tell you about is so absurdly abstract it's meaningless.' But before you know it, it just sort of blows up in your face." Arnold had one date who categorically refused serious involvements with badge holders. She feared the eyes of government would focus on her.[8]

Arnold's lost love may have been overly paranoid. Employees at Sandia seemed quite sensitive to security regulations, perhaps because the military base and the fence around Tech Area One were constant reminders of the division between the secret and nonsecret worlds. At Los Alamos people took a more cavalier attitude toward the rules. Too much focus on security and the laboratory might seem more like an armed camp than a scientific summer camp.

For example, the scientists in X-division at Los Alamos design the configurations for thermonuclear weapons. They work with top-secret data. Classified documents have red-and-white candy-

---

[8] As did I. I described my research in advance to the Public Affairs offices of both laboratories. Both of the laboratories' in-house newsletters kindly printed my request that people contact me if they were willing to be interviewed about their ex periences at the labs and their ideas on the social responsibility of scientists and engineers. A few people told me they had been asked to report back to the Public Affairs officers at Los Alamos and Sandia on their interviews with me. When I filed Freedom of Information Act requests with the departments of Energy and Defense and with the Secret Service and the FBI to find out if any checks had been done on my background, I was informed that there was no file in my name at any of those agencies.

striped borders that make them hard to miss. But when the stuff
in the documents seemed particularly interesting, X-division researchers would ignore the "need to know" rule and make photo-
copies to share with one another. This practice was so widespread
that one wit made up a rubber stamp reading "Bootleg," so the
real documents could be distinguished from the unauthorized
copies.

Part of the good-natured disdain for the security procedures at
Los Alamos stems from the scientists' skepticism about the competence of the guards. Many of the security guards at Los Alamos
are local Hispanics from the Española Valley, and guards usually
have no education beyond high school. The Ph.D.s' attitude of
superiority made it easy for them to joke that the armed sentries
posed more of a danger to themselves than to enemy agents.
When X-division suffered a rash of thefts—eighty dollars was stolen from a secretary's purse, and calculators and jackets regularly
vanished from people's offices—the newly instituted searches of
staff members' briefcases and purses turned up nothing. "Everyone
knows it's the guards," an X-division administrator said, shrugging.

Their cynicism may be warranted. A few years ago the laboratory administration asked anthropologists at the University of New
Mexico to design new procedures manuals for the guards. They
had in mind a comic-book format that could be understood by even
the most simple-minded security officer. Later, the lab administrators realized that secret security procedures would be described
by anthropologists without security clearances. They decided to do
the job in-house.

Disgruntled with their working conditions, in 1989 the 250
unionized guards went on a two-month strike against Mason and
Hanger-Silas Mason, Inc., the company that contracts with the
DOE for security services at Los Alamos. The company brought
in about sixty guards from other facilities and assigned them
and company administrators to twelve-hour shifts to replace the
striking security officers. When the DOE augmented that force
with forty guards from other security firms, the striking workers
cried foul, claiming the federal government was employing strike-
breakers. Meanwhile, the laboratory publicly denied any security
problem with depending on a reduced and exhausted guard force.

They could hardly do otherwise, but their sunny view of the situation might easily seem an official confirmation of staff members' suspicions that the guards were worthless.

When Gene worked as a technician at Los Alamos, he found secrecy a burden on his consciousness. "They use codes," he said. "For example, they had a project named White Horse, and they had a whole bunch of them named after horses. If you were talking to someone in a bar and they liked horses, you had to be really careful." You couldn't even use the phrase "White Horse," he explained.

Gene could mention White Horse because the project had been made public, as I discovered a week later when a Los Alamos physicist obligingly described the White Horse proposal: machines in orbit around the earth would shoot down Soviet missiles with beams of neutral hydrogen atoms. (Unlike an ionized or charged atom, whose trajectory can be bent by the force of magnetism, a neutral atom will keep on a straight path through a magnetic field.) Research on this weapons concept predated the Reagan administration, the physicist said. Fortunately for his group, White Horse became a Los Alamos favorite, its funding prospects brightening the moment Reagan embraced Star Wars. By mid-1989 Los Alamos was ready for the first SDI rocket test of a neutral particle beam in space, at the White Sands Missile Range. The White Horse had become a BEAR: Beam Experiment Aboard a Rocket.[9]

Secrets ebb and flow with the scientific and political tides. When one group of researchers at Los Alamos realized that the Soviets had stopped publishing experiments on a particular phenomenon, they knew it meant the Soviets saw the same potential weapons applications that they did—and the Los Alamos group's own research was promptly classified.

Sometimes these shifts in classification status were confusing, particularly for the older men. A thirty-four-year veteran of Sandia insisted that he could say almost nothing about one of his many engineering projects at the lab. From what he did say, I immedi-

---

[9]"Los Alamos Readies First Test for Star Wars," *Albuquerque Tribune*, June 10, 1989, p. A-3.

ately recognized that he had worked on nuclear power generators for satellites and rockets—the Space Nuclear Auxiliary Power systems. The project was described in an informative display in the Atomic Museum on Kirtland Base. "That sounds like SNAP," I said. Shocked, he sputtered, "How did you know that?" [10]

It stands to reason that New Mexico must have more spies per capita than any other state in the nation. In 1988 the Albuquerque Area Operations Office of DOE was responsible for nearly 47,000 people with Q-clearances. The Land of Enchantment, with an adult population of about 900,000, had roughly 40,000 residents with military or civilian security clearances. Someone has to be watching them.

Only one Los Alamos or Sandia employee has ever been caught spying, according to Jerry Brown, Sandia Laboratories' head of internal security. The spy was Klaus Fuchs, a Manhattan Project scientist who gave bomb secrets to the Russians after World War II. The data he passed along may have advanced the Soviets' own inevitably successful atomic bomb project by one-and-a-half to two-and-a-half years. Today, foreign agents might have a hard time recruiting confederates from the well-paid workers within the labs, since Brown's study of "insider" espionage throughout the defense complex over nearly four decades revealed that the American spies' overwhelming motivation was greed. With a track record marred only by Fuchs's ideologically motivated duplicity, it might seem that the labs do a pretty good job of keeping their secrets inside the fence. [11]

However, it has been relatively easy for outsiders to get *into*

---

[10] Within three years the space nuclear reactor program began grabbing headlines in the Albuquerque press. Its funding had more than tripled (with $143 million a year going to Sandia and Los Alamos in 1989), and a series of annual conferences in Albuquerque brought together researchers on the topic from around the world, including the Soviet Union. The tone was optimistic, except for a three-hour panel analysis of the public relations failures of earth-bound reactors. See Spohn, "The Force Is with Us," and "Space Scientists Fend Off Nuclear Safety Questions," *Albuquerque Tribune*, Jan. 24, 1989, p. D-1.

[11] Byron Spice, "Money, Love, Revenge Inspire Spies' Betrayals: Expert Studies Security Threats to Labs," *Albuquerque Journal*, Apr. 7, 1989, p. A-1.

the laboratories. In 1985 Los Alamos was shaken by the revelation that a former CIA agent, Ed Howard, had made several visits to the laboratory. Howard had left the CIA a bitter man—he was considered a poor agent and had been referred by his CIA bosses for psychiatric counseling. Forced to resign his job, he eventually became an economic analyst for the New Mexico State Legislature. He also sold secrets to the Soviets.

Howard, who had worked in the CIA's Moscow station, spilled the beans about the CIA's operations and data sources. The FBI bungled its stakeout of his Santa Fe home, and Howard eventually surfaced in Moscow, where presumably he will stay.[12]

Shortly thereafter, at the request of John Glenn, chair of the Senate Committee on Governmental Affairs, the U.S. General Accounting Office began an investigation of the extent to which foreign nationals participate in activities at the DOE weapons laboratories. The GAO found that between January 1986 and October 1987, Los Alamos, Sandia, and Lawrence Livermore had nearly seven thousand visitors. About fifteen percent of those visitors (just under nine hundred) came from either communist or "sensitive" countries. The sixty-eight "sensitive" countries suspected of developing nuclear weapons or viewed as posing some sort of national security risk include places as diverse as El Salvador, India, Israel, South Korea, and Vanuatu.

The weapons laboratories are also scientific and engineering labs, as employees take pains to explain, which is why it makes sense for so many outsiders to stop by to see what's cooking. But guests from communist and sensitive countries are supposed to be investigated in advance. The laboratories are supposed to request that the DOE have an "indices check" run by the FBI and CIA, which means that those two agencies check their files to ensure that a foreign scientist proposing a visit to the weapons laboratories is not linked to a foreign intelligence service. During their visit,

---

[12] Lance Gay, "N.M. Spy Snooped at Los Alamos Lab," Scripps Howard News Service, reprinted in *Albuquerque Tribune*, Oct. 22, 1985; David Wise, *The Spy Who Got Away*, (New York: Random House, 1988).

foreign researchers are not supposed to be briefed on any of the "sensitive subjects" identified as weapons related without prior approval from the DOE. And within a month after the visit, "host reports" describing the subjects discussed by the visitor should be submitted to the DOE.

By calling its report "Major Weaknesses in Foreign Visitor Controls at Weapons Laboratories," the General Accounting Office committed the sin of understatement. For example, the GAO reviewed the files of 116 people from communist countries on short-term visits and longer-term research assignments to Sandia and Los Alamos. Required background indices checks were done in advance for 3 of the 116. For nearly 90 percent of the visitors, the indices checks were *never* completed. Lab employees were barely more responsible about submitting host reports describing the nature of the visit: over 40 percent of the required reports were never filed.

Despite unexpected snags in their investigation (like CIA and DOE refusals to release some files), GAO auditors identified at least six individuals who "might have posed an unacceptable risk" but were allowed into the weapons laboratories without background checks. They also discovered thirty-seven visits covering sensitive subjects without prior DOE approval. [13]

Senator Glenn was incensed and the media had a field day. Laboratory officials explained over and over that just because someone gets inside the fence doesn't mean he or she gets to see any real *secrets*.

Kinks in the procedures caused the problems, according to the analysis by the congressional investigators. But there is more to the story than that. Once you know something about laboratory researchers' attitudes toward secrecy, their disregard for the rules and especially for the mounds of paperwork prescribed by the DOE makes sense. Laboratory scientists see secrecy as a boring

---

[13] U.S. General Accounting Office, GAO/RCED-89-31, *Major Weaknesses in Foreign Visitor Controls at Weapons Laboratories*, Oct. 1988; U.S. GAO Testimony Before the Senate Committee on Governmental Affairs, *DOE's Foreign Visitor Program Has Major Weaknesses*, Oct. 11, 1988.

and generally meaningless inconvenience. Many said that the decisions about what is classified defy all logic. When a scientist has traveled far to share his knowledge and learn from researchers at Los Alamos and Sandia, the enthusiastic researchers in their world-class laboratories are proud to hobnob with their fellow wizards. It has been easy enough to dismiss the rules and regulations set down by those obsessed with details, especially when the bean counters themselves are behind in their paperwork.

For most scientists inside the labs, the poorly educated guards in their spiffy uniforms are not the bogeymen. Other scientists, communist or not, hardly seem menacing either, especially in the heat of a discussion in their universal language about the surprising details of a new experiment. If they are afraid at all, weapons lab scientists are afraid of the spooks.

The spooks are the spies on *our* side: the plainclothes security officers, the efficient men with Q-sensitive clearances who interpret the intelligence information gathered about enemies and advise on policy, the wizards of technology who can read compromising emanations and trace them back to their source. If these people overhear you mumbling into your beer about white horses you will have a lot of questions to answer later.

In principle, no one with a Q-clearance should be uncomfortable answering the spooks' questions. Lab employees fill out annual activities reports (listing memberships and contributions to civic and political organizations, for example). In theory, the DOE updates the files for every security clearance every five years. In fact, the General Accounting Office found in 1987 that the DOE often failed to conduct accurate preemployment investigations before issuing Q-clearances. Despite regulations, Los Alamos and Sandia did not check to see if their potential employees had criminal records or bad credit, and Los Alamos frequently neglected to confirm employment and educational records.

Furthermore, the GAO found that the DOE was woefully behind in updating old clearances. By the end of 1988, the situation had improved somewhat: the DOE had developed reasonable plans for dealing with the backlog, and the Albuquerque DOE office had reduced the total number of clearances issued by 4,311. Only 53 percent of their clearances were more than five years old, so the local office had a backlog of only twenty-five thousand or so cases to reinvestigate.[14]

Researchers at the laboratories are caught between cultures. Science and engineering encourage dialogue. You learn by sharing information and ideas, and science does not work without openness and truth—lies and omissions are revealed by nature itself, as other scientists investigate the same topic. Researchers like talking about their work. I saw it happen again and again. Describing his job over coffee, a laboratory scientist would get enthused, reach for my pencil, and begin doodling diagrams, charts, and equations on a napkin to be sure I understood what their work was about. "The most beautiful thing about that one project I did," one man remembered, "is that I could *talk* about it."

The spooks want the scientists to make a habit of silence. Acclimating to that demand may simply involve accepting the already cosmic distance that separates scientists from laypersons. After all, communicating the meaning of a scientific discovery to an untrained audience is such a difficult task that scientists generally rely on journalists and popularizers to translate from the technical to the ordinary vocabulary. Silence around those who lack security clearances is essentially a logical extension of the intellectual isolation experienced by any specialist. Several Sandia employees felt the restrictions on them were identical to the constraints they had felt in the private sector, where they were responsible for protecting the proprietary secrets of profit-making engineering firms.

Still, the habit of silence takes on a life of its own. One evening I was talking with a man who complained that far too many pieces of information were classified. He proudly described his blow-by-blow battle against lab administrators obsessed with secrecy. He

[14]U.S. General Accounting Office, GAO/RCED-87-72, *DOE's Reinvestigation of Employees Has Not Been Timely,* March 1987; GAO/RCED-88-28, *DOE Needs a More Accurate and Efficient Security Clearance Program,* Dec. 1987; GAO/RCED-89-34, *DOE Actions to Improve the Personnel Clearance Program,* Nov. 1988.

had won, he said, and now the information on the number of neutron generators in a nuclear weapon was no longer classified.

"How many?" I asked. He did a double take and hesitated. He blushed and cleared his throat. Finally he mumbled, "Some have two."

8

ON THE SIDE
OF RIGHT

.
.
.

On August 6, 1985, the Santa Fe Peace Coalition planned a demonstration at Los Alamos National Laboratory to commemorate the fortieth anniversary of the bombing of Hiroshima. Group members had sought the necessary permission from the lab, which allowed them exactly three hours of public protest on lab property—from 7:30 to 8:30 A.M., at lunchtime, and again during the afternoon rush hour.

The officially designated protest area was across the highway from the administration complex, on a small grassy yard separating the road from a big laboratory employee parking lot. For anyone driving by, the protest area was impossible to miss. The lab had roped off two small triangles on either side of the paved path between parking lot and road with bright yellow plastic ribbons punctuated at intervals with small hot-pink streamers. That morning six journalists—including one representing a Japanese newspaper—showed up for the demonstration. They gingerly circled outside the largest of the two triangles. Inside stood three protesters.

Two uniformed guards loitered near the edge of the road. Laboratory workers rushed down the footpath toward the administration complex as an elderly woman prayed silently, her eyes closed. A thin blond woman with nose and eyelids burned raw by the sun practiced tai chi exercises while her dog rolled on his back and looked confused by the hurried foot traffic. With his fellow protesters occupied by prayer or exercise, the task of dealing with the reporters fell to a young man named Howard Shulman.

After getting their quotes the journalists, too, headed into the lab. Ed Grothus ran by, stopping barely long enough to hand each of us one of the colored postcards he'd had printed up for the anniversary of the bombing. On the front some tiny doomed palm trees were silhouetted against an orange-red mushroom cloud. The photo was from the Los Alamos Laboratory archives, it said on the back. And also this, next to the small rectangle labeled "Stamp":

> This is a product of Los Alamos National Laboratory. Weapons laboratories in other countries produce similar products. The potential of these vaporizing products is megadeath, ecocide and Cliocide (the End of History). Weapons workers are "good" people, "only following orders." YOU are responsible. Every person should do something to take away from world leaders this life threatening and despairable potential. ONE BOMB IS TOO MANY.

Apologizing for not staying to join the protest, Grothus explained that he was in a hurry to get a seat at the special public lecture by Harold Agnew, the lab's former director. Agnew was the man who had filmed the bombing of Hiroshima.

Previous demonstrations outside the laboratory had caused trouble, according to Shulman. The blond woman with the dog had been arrested in November of 1984 when she disobeyed local police and stepped off the sidewalk onto the street to hand leaflets to passing motorists. The year before, protesters had blocked Diamond Drive, another major road to the lab, and more than thirty were arrested, jailed, fined, and placed on twelve months' probation.

The Santa Fe group decided those actions were more alienating than constructive, and changed its tactics. It agreed to forswear street blockades, and a small group started showing up every two months with its members' thoughts on war, peace, and moral responsibility mimeographed on a two-page handout they titled the *Cottonwood Truth Relay.* (Like the Los Alamos aspens, cottonwoods are poplars. Cottonwood groves line the Rio Grande Valley.) Members of the Santa Fe Peace Coalition thought truth could be carried up the Hill. But when they tried to leave copies of the *Cottonwood Truth Relay* in the laboratory's Oppenheimer Memorial Study Center, the head librarian had refused them. Shulman said he felt sorry for the guy, he'd been so obviously embarrassed. "He said he was *told* that he didn't want them," Shulman said. Shulman thought that the decision came down from administrators responsible for laboratory security.

"Check that out," Shulman said, inclining his head toward the fire station a few hundred yards down the road. A man was on the roof, doubled over a tripod and what looked to be a camera. "He set up at 7:30 when we arrived." Shulman guessed he was aiming some sort of directional microphone at the protesters on the grass. He turned and waved to the photographer—no response. Then he pointed across the street to the Otowi building, where the lab's personnel office is located. Five men stood in a line along the smoked glass wall of the cafeteria on the second floor.

"They're the security force," he explained. "They always watch us like that." Behind us three fire engines pulled out onto the driveway of the fire station, ready for anything.

All that happened though, was the arrival of another young man who had driven down from Taos, New Mexico's famous ski resort and artists' colony. Pete brought four huge silk-screened banners that hung vertically from bamboo crossbars. His heraldic emblems of the nuclear age showed the skeletal remains of the round dome of Hiroshima's Hall of Industry beneath a flock of doves. By that time foot and car traffic had thinned to almost nothing. A young Hispanic woman stopped on the path to gape. "I like your signs," she said finally. "How much do they cost?" "Why don't you go home?" someone shouted from a passing pickup truck.

At precisely 8:30, everyone—demonstrators, security men, and the rooftop observer—packed up and disappeared.

I wandered toward the cafeteria. In the nearly deserted halls of the Otowi building I ran into a man I had interviewed a few months earlier. "I'm here to watch the demonstrations," I explained.

"Oh *yes*," he said, light dawning in his eyes, "today's the sixth. Are you going downtown to see the parade later?"

"What parade?" Could they have a *parade*?

"You know, down by Ashley Pond, the parade," he said with an innocent look.

"Parade?" I repeated, dumbfounded.

He shook his head, the innocence on his face giving way to disappointment. "No, no, it's just a joke. Relax. There's no parade."

I felt like a fool. But I think it fair to say that there was unusual tension in the air that day. Many lab employees had smiled at the protesters on their way to Agnew's lecture that morning, but the smiles were tight and there was a good number of glares and grim looks. A week of television documentaries and newspaper headlines about the birth of the atomic age had taken its toll. When I left Albuquerque at dawn for the long drive to Los Alamos I startled two men furtively spray painting a human shadow in black on the corner sidewalk outside my house. As I watched, one stenciled the explanatory slogan beneath it: "Hiroshima—Never Again." The shadow project was nationwide, he explained over the hiss of the spray can. The early-morning airwaves buzzed with stories about the fortieth anniversary. It seemed unlikely that people in Los Alamos had forgotten the significance of the sixth of August.

I had just decided to go into town when Agnew's lecture let out. On my way to the parking lot I ran into Sigmund, a young West German mathematician on a two-year visitor's appointment to the laboratory. We had met before, but Sigmund barely recognized me. He was wild-eyed. "I have to get away from here, let's go, please let's leave right away," he said, barely breaking stride. I drove, and we ended up at a small coffee shop just off Trinity Drive.

Sigmund was enamored of the American West and had adopted as his regular costume old blue jeans, scuffed cowboy boots, and a white ten-gallon hat. Along with his unkempt brown hair, thin beard, and granny glasses, this outfit made him look like a shy young European intellectual in pursuit of rugged individualism. Sigmund had come to Los Alamos to study with other mathematicians in the famous T-division. But he had mixed feelings about the laboratory.

Sigmund had been to meetings of the Santa Fe Peace Coalition. As far as I know, he was the only person currently employed by the lab who had demonstrated for peace outside the Los Alamos National Laboratory. He wanted people at the lab to think about what they were doing, he explained.

In West Germany, Sigmund probably would be considered close to the center of the political mainstream. In Los Alamos, where voters register two-to-one Democrat and vote two-to-one Republican, he was out in left field. Earlier he told me that he found it depressing to be around so many smart people doing such a stupid thing—"The bomb, I mean." But that morning he was nearly beyond words.

"The name of Agnew's talk was 'The Way It Was: No Regrets,'" Sigmund mumbled. So, he explained, he wasn't that surprised by the overall tone. But his voice cracked when he repeated Agnew's closing line. "'They bloody well deserved it.' Those are the words he ended with. And they all applauded."

Sigmund was inconsolable.

When we returned to the lab an hour later, we walked together from the parking lot to the Otowi building. Sigmund continued on to his office ("I'll be okay," he said), and I went upstairs to the cafeteria to kill time before the lunchtime demonstration.

Shortly before 11:30 four men took over a table beside the glass wall overlooking the firehouse and the parking lot. I watched them for a few minutes: a blond in his forties, two younger dark-haired men, one in the khaki jumpsuit and high black boots of the laboratory's Rapid Response Team, and a very heavy black man. They looked like observers from the security force and they looked bored, so I went over, introduced myself, and asked to join them. Everyone stared. The black man said, "Sure, sit down."

For half an hour I watched them watch the protesters reassemble below. The young fellow in the khaki uniform had a Texas accent and an Elvis hairstyle. He was the only one officially on duty. The others were just keeping him company, they said. Along with the radio and weapons strapped to his uniform was a pair of binoculars that passed from hand to hand around the table.

When it was my turn I could see that the original three demonstrators had been joined by four others, the dog still looked confused, and the photographer on the roof of the fire station had not reappeared. "What photographer?" one of the men said. "I didn't see any photographer." Everyone swore that they hadn't noticed anyone on the roof of the firehouse that morning. "Maybe he was a naturalist—you know, photographing the forest," one said. "Or a highway surveyor," another added helpfully.

The big black man had a face that split with a smile at the slightest provocation. He was a security supervisor, he said. I told him I had been interviewing people at the lab to find out how they felt about their work. "Well, let me ask you this," he said, leaning forward in his chair. "Do you believe in God?"

• • •

Patriotism is an old-fashioned virtue. As the world gets smaller and social institutions proliferate, the idea of a natural obligation to obey and defend anything seems increasingly irrational and quaint. It is one thing to go along with national policy because that policy seems right, quite another to comply simply because it is policy.

Whatever their own feelings, most people at the weapons laboratories thought their coworkers in the laboratories were motivated by love of country. I had interviewed Seymour within a few days of the thirty-ninth anniversary of the Hiroshima bombing, a day completely unnoted in the press. All of his physics research at Los Alamos was weapons related, Seymour told me, barely getting the words out between bites of a huge Spanish omelette.

The Los Alamos Lab cafeteria has excellent food. After we had worked our way through the line and settled down to talk, Seymour began the interview by posing his own question: "What motivates people?" He meant, "What motivates people to work in a weapons laboratory?"

Seymour pursued the answer in a hurried imitation of Sherlock Holmes. First he eliminated false leads, like brains and money. As a graduate assistant, he had struggled to teach elementary physics to premed students. He figured that physicists like himself were smarter than those kids destined to earn $100,000 a year—smart enough to do whatever they wanted and obviously not greedy enough to plan a lucrative professional career. Job security couldn't be the motivation, Seymour decided, because tenure at a university is theoretically more secure than a contract at the laboratories. Because of the laboratories' secrecy, professional accolades could be eliminated as a possible motive. Ergo, it must be patriotism.

"It's one of those necessary things that we need to do to preserve the values that we think are important," Seymour said, slopping a flour tortilla in the red sauce left on his plate.

Seymour had served in public office in the county and was up for reelection. He put his own quotes around his clichés about political ideals, interrupting himself in the middle of sentences to say things like, "Now see, this is campaign stuff." So what values motivated *him* to work at the weapons labs?, I asked. "Well," Seymour grinned, the quotation marks implicit, "we work here to preserve the ideals expounded in the Constitution and the Declaration of Independence."

"You can easily look backward and say that the guys who worked here in the early forties did the right thing," Seymour said later. "There has never been a World War III. With a tiny fraction of the world's resources, we have prevented this mass destruction. But of course," he added cheerfully, "you can't look forward and see if it will hold in the future."

Seymour was happy with his job. He worked in a "sort of amusing office" where people talked a lot, and a lot of that talk was about politics. His group did some of the technical analyses used by government officials considering the implications of arms limitations treaties. But Seymour had to imagine the basic motivations of his coworkers. The ethical dimensions of their own work was never a topic of conversation.

"Everybody, I think, has a somewhat similar view," Seymour explained. "Or else you wouldn't be here. What are you going to do, wring your hands?" Seymour smiled, exuding goodwill and self-confidence. "I don't think of agonizing about my work," he continued. But when he had finished his breakfast and rose, smiling again, to shake my hand and wish me luck with my next interview, he added, "Maybe your next interview won't be such a cop-out."

I did a double take and asked him what he meant. Seymour just kept grinning and repeated the comment as if I must have misheard him. "What do you mean?" I asked again. Seymour shook his head. He said he figured I must have expected to meet people who were into wringing their hands. He, personally, did not feel that way. Still smiling, he waved me away.

•   •   •

Others at the labs were unself-conscious about saying that they work for the defense of the United States. Eloy lived in the Española Valley below Los Alamos and had been at the laboratory for twenty years. His family had inhabited the highlands of northern New Mexico and southern Colorado for as long as anyone could remember.

He led me into the kitchen to meet his wife. Gloria had caught a chill in the yard and huddled shivering on a dinette chair in front of their woodburning stove. The stove was in a kitchen corner, the wall behind it covered with aluminum foil. "Reflects the heat so it saves wood," Eloy explained on his way into their small wood-paneled family room. Gloria followed us, grabbing a knitted afghan off the back of the couch.

Eloy served in the navy after World War II. There he learned about bigotry against Hispanics. While playing the piano in a military recreation hall, he felt a man come up beside him. "He said he didn't know that Mexicans could be taught to play the piano," Eloy said incredulously.

"I don't like when they call us Mexicans, you know?" Gloria

interrupted from inside her afgan cocoon on the couch. "Like it's an insult. Like we're dirty."

Gloria worked as a secretary for the Zia Company, the firm that until recently contracted security and maintenance services to the lab. She and Eloy glanced at each other and on some unspoken cue gingerly began talking about racial prejudice at the lab.

The founders of Los Alamos Laboratory chose their site because they saw the high mesa as isolated. But that wasn't true, as Chris Dietz had pointed out. The people discounted by the Manhattan Project scientists were poor Indians and rural Hispanics.

"The Los Alamos people really look down on Española, you know that?" Eloy said. Forty years after the laboratory was founded, "Española jokes" were all the rage.

What is the definition of Father's Day in Española? Chaos and confusion. Why did the Española zoo close down? The chicken died. Some said people laughed too hard. Others said relax, it's a joke. The Zia Company (and its successor) employed a lot of Hispanic workers from the Española Valley. The company issued a policy memo threatening disciplinary action against any worker who persisted in telling ethnic, racial, or sexual jokes, including attacks on Española.

"What the lab has done recently has been to hire some very professional people who try to—well, they squelch the Hispanics before it gets out of hand. They say, 'Don't go to court, go to the affirmative action officer,' but really they're trying to squelch it," Eloy complained. "Look, I served my country. No one questioned me then. No one asked was I Hispanic. I was proud to serve my country.

"I was a crypto man and I was able to see communications from many parts of the world," he continued. "So I have an idea of what's going on. From a moral standpoint, I think we have to stand strong, because we are competing with an evil government, a totalitarian government, which is the Soviet Union.

"People are under the wrong impression that if we were to disarm this country, the Soviet Union would say, 'Okay, we'll disarm and we'll live in peace' and all that. They would rather enslave us," Eloy said. "Have you read *The Communist Manifesto*? I think that lays out their plan pretty well."

Eloy had no delusions about the effect of nuclear war. "I had the experience of being at Hiroshima and, uh, and the other city that was bombed." He and Gloria both looked embarrassed that Nagasaki had slipped his mind. "I helped make the initial measurements at Eniwetok. If anybody has a sense of what the atomic bomb is like, I did. Hiroshima was completely wiped out." But then, Eloy added, you should think about how many American lives would have been lost in the planned invasion of Japan.

Every Sunday Eloy and Gloria went to Mass at the Catholic church across the road. Eloy felt there will be a major nuclear war. He thought the only solution would be the total integration of the human species, but he saw no way to accomplish the unification of humanity short of disaster. The Book of Revelation talks about two armies, the good and the bad, Eloy reminded me. "All these prophecies have to be fulfilled," he said as Gloria nodded. "And I think it's going to be soon."

● ● ●

Henry was a sixty-three-year-old moon-faced Sandia engineer who read the Bible for inspiration and comfort. He limped slightly as he made his way, cane in hand, across an acre of light-blue living room rug and around a yapping curly-haired puppy to the dining room table.

"I was in a POW camp for two-and-a-half years," Henry said. "I lost a leg when I was captured, so when I got out I spent a year and a half in a hospital." A *Sandia Lab News* story about his experiences said that he'd flown thirty-eight missions as a fighter squadron commander before the North Koreans shot him down and took him captive in 1951. Henry survived a crude amputation, six primitive operations on his good leg, and a 140-mile forced march. He mentioned none of these things when I asked him about his twenty-seven-month ordeal. But he knew I'd heard he had cancer, and admitted that his chemotherapy made him tire easily.

Henry summed up his view of communists in one word: aggressive. "If you start reading their books—*Das Kapital* and the works by Lenin—you see that it's not just the leaders that are aggressive," he said. "For communism to succeed, it has to conquer the world under one dictator."

I must have looked skeptical.

"You would be a more likely target for their wrath than I would," Henry insisted quietly. "You have a social position. As an engineer, they could possibly use me. But you, a sociologist, they'd have to change your mind or do away with you."

Henry wasn't much of a talker. After each statement he would close his mouth firmly and wait for the next question, his eyes wandering between his hands on the table before him and the puppy bouncing off the living room furniture and running tight, enthusiastic circles on the blue rug. I asked what it was like in the prison camp.

"We were subjected to daily doses—like ten hours a day—of communist philosophy and propaganda," Henry replied. "I wished I'd taken more social science at West Point, because I needed to have answers." The program was drummed into him. First, revolution. Then the dictatorship of the proletariat. Finally, the state withers away, replaced by the eternal utopia of true communism. Because Henry was a lieutenant colonel, he was besieged by confused officers and enlisted men seeking reassurances that the Marxist plan was bankrupt. But the U.S. military in the early 1940s had been preoccupied by Hitler running amok in Europe, and West Point had given Henry no ammunition for fighting an ideological cold war.

Henry did the best he could. The citation accompanying his Distinguished Service Award was reprinted alongside the essay he'd written for the *Sandia Lab News*. It credited him with refusing to give up secret information despite sadistic torture and incessant interrogations, and with heroic self-sacrifice for the sake of his fellow prisoners.

When the North Koreans and the United Nations forces agreed to exchange their sick and wounded prisoners, Henry was taken from the camp and put alone in a room. He heard an official announce fictitious details of his reunion with his family over the camp loudspeaker. The lie was his final lesson on communist morality.

"You know, being in a prison camp and then a hospital is a very boring life," Henry said. "When I got my retirement papers from the Air Force, I just started heading west." In Albuquerque he saw a sign advertising an employment office, and Sandia Laboratories received his first and last job application. The lab had treated him well. Henry did not become a weapons engineer as a public service in defense of democracy. Instead, he said he had liked his job for the challenge of the work itself—work he could do with one leg.

"Having lost a leg, that kind of gives you a bad attitude," Henry said. "I'm certainly well off financially. I don't have three Cadillacs in the garage, but I'm doing pretty well. Losing a leg, you feel like you can't do as much, achieve as much, and you don't expect as much."

"I've been through two wars. I faced down a Japanese fighter plane at thirty thousand feet with him firing his cannons and me firing my machine guns and both of us hoping the other would break away first. If we didn't have nuclear weapons we'd have something else, poison gas or chemical weapons or something."

Like Eloy, Henry had seen Hiroshima and Nagasaki after they were destroyed. "A nuclear weapon is a pretty horrible thing," he said. "But I also saw napalm, and I saw Tokyo after it was bombed, and at the time I didn't think nuclear weapons were any worse than conventional weapons." He hesitated. "Today they are worse. They're more powerful. I'd certainly like to see a reduction in the numbers of them, particularly some of the big dirty ones the Russians have."

"Dirty" weapons produce a lot of fallout and radiation. "Clean" bombs have fewer collateral effects. Henry had worked for twenty years on safety procedures and devices for nuclear weapons and the nuclear components of rockets. The American emphasis on nuclear safety and safeguards research served as proof of America's superior morality, he thought. His group at Sandia used to dream up scenarios, Henry recalled, like, "On the launch pad, what if you had a rocket sitting in 160-mile-per-hour winds and it toppled? You could have a maximum explosion. How could you protect the people of Florida from this type of catastrophe? It could very well put down a path of radioactive contamination fifty miles wide and three hundred miles long, depending on which way the wind was blowing. We were hoping it would blow toward Cuba, if it happened at all."[1]

Henry believed that the Soviets placed a much lower priority on protecting even their own people. "Human life doesn't have much value to them." The puppy was doing gymnastics around the leg of his chair, and Henry gently scooted him away with his cane.

Antinuclear protesters struck him as sincere but misguided. "They're kind of in the category of the environmentalists who make it necessary to get an environmental impact statement before they can grade a road in the county so some Indian kids can get to school," he said, almost but not quite smiling. But Henry felt he understood the goals of both the communists and the peace activists. They dream of peace, brotherhood, and sharing, as did he. Henry found those same ideals in Christ's Sermon on the Mount. The end we all long for is the millenium, Henry agreed. The only real issue was the appropriate means.

•   •   •

Back in the cafeteria at Los Alamos, one of the security men left the table when the topic of God came up. He emerged at the base of the building and headed across the street toward the protesters.

"Do you believe in God?" The black security officer persisted.

"Do you?" I responded. This did no good. He rejected the diversion and shook a fat finger at me.

"Now I'm asking *you*, do *you* believe in God?"

I said no.

---

[1] We spoke before the space shuttle *Challenger* exploded during launching. Some government officials subsequently raised questions in the *Challenger* case about precisely the sorts of contingencies he had been analyzing: the disposition of radioactive materials within a rocket, or its payload, in case of accident.

The man from the Rapid Response Team looked up, surprised. The supervisor smiled broadly and settled himself firmly in his chair.

"Let me ask you this. Do you think the sun will come up to-morrow morning?"

I did, and he was ready with a series of questions designed to turn the atheist. How did I know the sun would come up in the morning? What created good and evil? Didn't all this require a Creator? Didn't I love life?

The jump-suited young Texan from the Rapid Response Team had edged closer, lips parted, poised to advance his own argu-ment: I said I loved life, right? Right. And I was alive, right? So I must love myself, right? Right. So if I loved myself, why was I smoking a cigarette? Through the tinted glass window behind him I could see the middle-aged white security man standing on the sidewalk between the yellow triangles, talking to Howard Shulman.

He was reminding them of the rules, the black supervisor ex-plained, heading back to the arguments for the existence of God.

I asked why my feelings about God were important. "It tells me what your bias is," he said promptly. The guy from the Rapid Re-sponse Team was reluctant to let go of his cigarette argument. "How about it?" he demanded, pointing to the ashtray. He wanted to talk about God some more. The black man looked at me side-ways and rolled his eyes.

God was interesting, I told them, but right then I was more interested in what was going on down next to the fire station. What did they think about those protesters, anyway? The young Texan seemed disappointed by the official shift of topic. His face went blank. Then he thought of something.

"Say, who was that I saw you walking with before?" he asked casually, his eyes narrowing. "That guy you walked with when you came over here before? You know that guy? That guy a friend of yours?"

He meant Sigmund. The black supervisor watched quietly, his smile a bit less broad.

"Yes, I know him," I said.

"Is he a friend of yours from the university?" the Texan asked slyly.

I told him no, that Sigmund worked at the lab.

"I know it!" the Texan exclaimed, his excitement restored. "I know it because I followed him to his office." He followed all the protesters, he said. The black man tucked in his chin. "And a major proportion of them work right here at the labs!" the Texan concluded triumphantly. Below us there were now seven protesters, tiny figures milling around inside a yellow triangle on the lawn.

•   •   •

People at the weapons laboratories were articulate about their work. Despite security restrictions, they loved explaining their scientific research. They explained things I have forgotten, producing crude little sketches that I can no longer decipher: Two ways of studying fusion, magnetic and inertial confinement. Symmetry in nature and mathematics. How lasers work. The derivations of the equations for the law of angular momentum, written in tiny letters on the edges of a soggy paper placemat. The General and Special Theories of Relativity. Linear and nonlinear equations. How an accelerator works, the design of monitoring devices for geothermal research, and a conceptual picture of stability in the arms race. Some, more facile than others, more quickly conceived a metaphor or illustration to help convey their point. But everyone who started talking about their research did so with visible enthusiasm.

They were noticeably less articulate about politics. Words seemed to fail the avowed patriots at the weapons laboratory. "To some extent I believe in the inherent goodness of our country," said one young Sandia engineer. "And to some extent I believe that we are an essentially peaceful, nonexpansionistic, right-thinking country. But like the Soviets, we sometimes become a little paranoid."

Scientists and engineers should apply their work for the benefit of society, he thought.

And what does that mean?

"I guess—war is wrong, peace is right. Life is right, death is wrong. Truth, liberty, and the pursuit of happiness is right. Tyranny, blah, blah, blah, is wrong."

Blah, blah, blah?

Why did employees of the weapons labs so often express their patriotism in the hackneyed phrases that accompany the tunes played by the blaring brass and tinkling cymbals of an all-American marching band? Why did they deliberately and self-consciously fall back on platitudes like "We are the good guys" who are protecting "truth, justice, and the American way," "the pursuit of happiness," and "the ideals expounded in the Constitution and the Declaration of Independence"?

Partly, I think, it is because some had stuck with the science courses in college. They were vague on the details of world history and perhaps embarrassed to venture beyond the slogans that had been anchored in their minds at least since their junior high school civics classes. Many took the long-dead Stalin as the exemplar of communist philosophy in action. Americans were the white hats, they said, and the Soviets were still and always Ivan the Terrible. In the context of the much broader current national debate on the future of international relations, stirring rhetoric from the cold war rings hollow when not backed by factual details.

People normally find it easiest to describe the things they know and love. For most people at the weapons labs, that meant their latest scientific problem or engineering puzzle, not the latest twist in U.S.–Soviet relations. But even the strongest patriots at the labs had trouble explaining what they knew and loved about the United States. They expressed their doubts more clearly than their certainties.

A woman engineer at Los Alamos figured we were trapped in the arms race. "The Russians are doing it full tilt. The Chinese are going full tilt. Once I got past the fact that other countries are doing it full tilt, but don't have the Christian ethics I have, I thought it was okay. And people who work on weapons better be damn good at it." Since *she* was damn good at it, she overcame her reservations and went to work in the weapons engineering division at Los Alamos.

"I'm Attila the Hun in an Itty-Bitty Body," she joked, explaining her conservative Republican outlook. She thought the Democrats in Congress who supported arms limitations foolish. "And I don't believe in welfare, I think those clowns ought to work. Or let them starve, I don't care." Her interpretation of Christian morality emphasized industriousness, required that you treat others as you would be treated yourself, and, above all, forbade justifying immoral means by reference to desirable ends.

"When I first came to the lab it worried me a little bit morally because what I was working on could kill people," she said. "But if bombs are not made right, they'll kill people anyway. And it has to be done, so *I* might as well do it so it will be done right." She had been shuffling papers on her desk as she spoke and stopped, manila folder in hand. "That's probably a cop-out," she said. "I remember thinking, 'This is a cop-out.'"

Did she see conflicts between her work and her Christian principles? She paused, then sighed. "No," she said. "But maybe that's because I don't think about them real deeply."

• • •

Two of the newcomers to the noontime protest on the fortieth anniversary of the bombing of Hiroshima were standing apart from the other five inside the yellow triangle. He was an Episcopal priest from Rhode Island in his mid-forties, wearing blue jeans and a clerical collar. She was about ten years younger, a seminary student from California.

They were on a cross-country camping trip, he explained, and they felt morally bound to detour through Los Alamos so they could demonstrate their complete and utter opposition to everything that went on in the weapons laboratory. They were eager to explain their views. "How could anyone claim to be moral and work on bombs?" they asked. I tried to explain what I understood of the attitudes of people inside the labs. They were too excited and too absolute in their own position to pay much attention.

"We're going to perform the Mass," the priest announced. The seminary student nodded and started pulling the necessary paraphernalia out of her shoulder bag. They had prepared a special prayer for the occasion, she explained, handing a folded piece of paper to her companion.

The other protesters were chatting among themselves. I told Shulman that I'd been delayed by my conversation with the guards in the cafeteria. "Well, you missed two positive honks," he reported.

The lunchtime traffic in Los Alamos is light. Most lab workers take a bag lunch or walk to the cafeteria. The priest read his prayer loudly, aiming at the scattering of cars that zoomed down the road between the protest area and the administration complex.

"And have mercy on us," he intoned at the end of the blessing ceremony, waving his upraised arms in the general direction of the administration building, "and on those who work in this place." At just that moment a hand stuck out the driver's side window of a passing car signaled with index and middle fingers a "V" for peace, or maybe for victory.

Eyes focused heavenward, the priest and his assistant missed the sign.

9

HOLY WINDS

•
•
•

A desert wind will cloud your eyes, drive grit between your teeth, and scour your skin. A sudden spring gust can make you stop in your tracks and turn your head. A maelstrom in the air can blow you off your feet. No wonder, then, that the traditional Navajo believe that forces of good and evil are manifested on earth as winds. The spirits swirl around you, urging you to shift direction.

Walk this way, one wind says. Another pushes you off track. The ideal is a balanced and appropriate path through life. When someone is in harmony with the natural order, the Navajo say, "He walks in beauty."

Walking in beauty is no easy thing. The currents of air circle and gust in unpredictable ways. A dust devil may appear out of nowhere. A steady wind gently encourages you to change your plans. Forces of nature are sometimes so subtle they cannot be perceived. The winds are present even when not felt.

How do you know when the winds are blowing you off course— off the "corn pollen path," as they say? How do you know when to resist and when to turn aside? As natural beings, people also have spiritual forces. We are all accompanied throughout life by two small winds called *nit ch'i yazhe*. From their perches in the shells of our ears they breathe advice, good to the right, bad to the left. When the left side spirit blows strong, people get into trouble. To straighten a crooked path we must sometimes muster the courage and strength of our larger spirits, which live deep inside us. These spirits—*nit ch'i iishii zinin*—rise from within to help us resist the temptations of the evil wind.

Early Catholic missionaries to the Southwest listened to the wind stories and believed they recognized a heathen version of the Judeo-Christian conscience. They used the stories to make Christianity more compelling to the desert people, interpreting the omnipresent natural wind spirits as the still, small voice of God directing the individual to resist blustery Satanic urgings.

But the Navajo friend who told me wind stories was displeased when I suggested an analogy to the internalized individual conscience. We were sitting on a grassy hillside in an Albuquerque park. He was describing how people go wrong and what can help put them right. When he talked about the winds he moved his hands in broad gestures, the slow loops and curves echoing the lay of the land and the turbulence of the wind spirits. When I said, "Angel on one shoulder, devil on the other," he shook his head and folded his hands on his lap.[1]

• • •

There are five major entrances to Kirtland Air Force Base, so Chuck Hosking follows a formal schedule to be sure he presents himself twice a week at each gate—once in the very early morning, when people drive in to work, and again in the evening, as they head home.

"Hiroshima-Nagasaki Forgive Us." "Trust in Weapons Leads

---

[1] For more wind stories and an analysis that disputes the analogy to Christian conscience see James McNeley, *Holy Wind in Navajo Philosophy* (Tucson: University of Arizona Press, 1981).

to . . . ?" "Be Realistic—If We Continue to Build Bombs, They Will Be Used." "Who Will Stop the Arms Race?" Hosking has twenty different messages for people employed on the base. His double-sided signs are painted on old bed sheets cut in half lengthwise, with the sides seamed up so each banner can be slipped over two six-foot poles. If someone joins him in protest, each holds one of the poles. When he is alone Hosking will lash one pole to a fence or signpost, or stake it to the ground and lean it against his old red bicycle.

The first time I met Hosking he was alone and had tied the pole to the chain link fence outside the gate at Wyoming Boulevard, one of Kirtland's busier entrances. The wind was blowing in sporadic but vicious bursts, roaring through Tijeras Canyon with typical springtime ferocity. Accordingly, the banner would sway unpredictably, fill with air, threaten to sail away, then hang slack. Hosking is in his mid-thirties, slender, with a dark tan and a shock of wavy black hair styled like John Kennedy's. That afternoon he was in green: stained dark green cotton shorts, a worn green plaid shirt, a green baseball cap, and thick-soled workingman's shoes.

At the Wyoming Gate to the Air Force Base a very long block of nothing stretches between the guard shack and the gas station where I had parked my car. With the sun in our eyes and the wind in our throats, Hosking and I were the only things standing still in the stream of cars and bicycles.

I have since stood with him at each of the major gates to the base, on a morning when it was cold and wet and on still summer afternoons when the heat and car exhaust seemed to conspire against breathing, but that first time was the worst. Everyone was driving out. We were facing in. The rhythm of the traffic was hypnotic, and the hostility was tangible. "I've gotten rammed into a couple of times by bicycles at this gate," Hosking said.

"Just yesterday two soldiers in a pickup truck sat watching me for about twenty minutes. Eventually, one of the guys came along and clobbered the banner." He demonstrated what had happened while I held the pole: the soldier raised both hands, clenched together, over his head and brought his fists down across the middle of the bed sheet. Hosking grinned.

"He broke the ropes—not a word, he didn't say a word. One time a couple of guys threw huge rocks at me and chased me on my bike for about a mile. Some construction workers started shooting at me with a BB gun. I got hit in the leg; it swelled up for a couple of weeks. And people throw glass at me, bottles and rocks. I used to tie the sign to the speed limit sign, but the police fined me fifteen dollars. The rent was too high, so now I stake it down, unless it's too windy, and then I tie it to the fence, like today."

Hosking did not hold down a job in the traditional sense. Squinting into the sunset so he could make eye contact with the blurred faces behind the windshields, he explained that his work was peace and justice. He calls what he does—standing outside the gates with slogans on banners—the Sandia-Kirtland Peace Values Project. He started on Ash Wednesday, 1983, and has been at it steadily, with and without companions, ever since.

This kind of work earns him no pay, of course. Hosking has a master's degree in history. Denied a social science teaching certificate because he had an arrest record (for protesting at the Pentagon against the Vietnam War), he taught mathematics in community colleges and alternative high schools for twelve years. Eventually he and his wife Mary Ann saved enough money to buy a tiny house in the poorer part of town, down in the South Valley along the Rio Grande. They can live on their savings because they buy almost nothing. Frugality and simplicity are moral imperatives for them. Hosking believes that true and lasting peace cannot be produced in weapons laboratories. Peace requires justice on a global scale, and justice requires a redistribution of the world's resources. "The issues are all connected," he says, "peace, justice, and affluence." By living poor, he can do his bit for redistribution and can afford to work as a full-time peace activist.

Hosking kept a record of the first year of the Sandia-Kirtland Peace Values Project. His log is written on the backs of flyers announcing other projects and events. It includes estimates of the amount of traffic at each of the gates in the mornings (6:30 to 8:15) and afternoons (3:30 to 5:15), along with a detailed breakdown of the rate of flow (as many as forty-three cars and five bikes per minute at the Wyoming Gate, for example, for a total of between three and four thousand vehicles each shift). For each vigil, he recorded the names of any others who joined him that day, the messages he had displayed, and a few comments:

After nine days, "arousing negative emotions, finally." On the eighteenth day, "rain and wind, cyclists getting more angry." Three months later, with the first appearance of the banner "Would Jesus Work in This Death Factory?", "unanimous: YES—lots of laughing (nervous?)" Two guys shouted "Getting tough," Hosking noted in small, clear handwriting.

It is rumored among Sandia employees that Hosking is an ordained minister. In fact he was raised Episcopalian and wanted to be a minister. But in the early 1970's Hosking's orientation toward peace and justice, originally inspired by the civil rights movement, seemed too controversial and worldly for his church, and he was not given the chance to be a priest. His God is one of peace and justice, a political savior.

Leafing through the pages of his diary, where each day gets only one line, is like looking at the plot outline of a very slow play. A little note in the log commemorated the day that Eli, a Vietnam-era vet who was holding the other side of the banner, was hit by a bicyclist. After four months someone asked, "You still at this shit, man?" One day Hosking counted 125 responses (when his sign asked "Will Your Kids Survive Your Work?"). The next day, with the same message, he got only a dozen.

Surprisingly, Hosking's abbreviated diary of his first year outside the gates of Sandia turns into a cliff-hanger. Day by day, line by line, you can watch the relationship form between Hosking and the cars, as people begin to wonder if he intends to stand there forever. A few prepare signs to press against the passenger window in rebuttal: "Forgive Hitler, too," and "USSR loves you." But more shows of agreement also appear: smiles, headlights flashed in solidarity, friendly waves, a man who stops his car and says, "I want you to know I'm with you." One man nearly caused an accident when curious drivers slowed to watch him walk over to drop off a long letter, the first in a series of exchanges between the two. I met the man later. He said as he drove past he always waved to Hosking and got no acknowledgment. I found it hard to see people's faces through the windshields.

Eventually the entries chronicle the onset of violence and bad weather, noted briefly as "drizzle," "cold rain," "motorist flings cigarette at me," "near zero degrees plus wind," "driver throws exploding firecracker at me," "motorist throws bottle, shatters at my feet." On the last page, on what he originally intended to be the last day of vigil, Hosking offered some summary statistics on the number of person-hours spent at the gate and the number of cars and bicycles that passed him. And he noted that the responses were fairly consistently negative, by a margin of three to one, regardless of the banner he displayed.

As I stood with him outside the Wyoming Gate that first afternoon, a grim-faced bicyclist broke from the pack on the sidewalk and swerved toward me, shouting something lost on the wind. Hosking thought it was a demand that we move over and leave more space on the sidewalk.

"It's the same mentality that destroys the world thirty times," he mused, still smiling at the traffic. "There's more than enough room there. Two bikes could easily get by side by side, but they want even more room in the same way they want even more bombs."

His unpaid regular work fills a void, Hosking thinks. He knows that the weapons are built because of policy dictated from above. "But if nobody would have operated Hitler's gas chambers," he adds, "no one would have been exterminated." And does he think that the Sandia-Kirtland Peace Values Project makes a difference? Hosking claimed no illusions. He wondered aloud if maybe he didn't do as much harm as good, by making people first defensive and then hardened against challenges to what he called their "armadillo armor."

"I do this because I feel I have to, not because I expect to be effective," he said. The traffic had slowed. "Is it 5:15 yet?" He pulled a watch out of his pants pocket. "Oh, it's past that." Without missing a beat, he rolled the banner around the poles, lashed it along the frame of his bicycle (which makes it difficult for him to turn the bike, a fact that worried one Sandia engineer), waved good-bye, and rode off into the sunset.

"This is the kind of stuff the churches should be doing. But they don't have the guts," Hosking said one brutally hot July afternoon

when his sign read "Would Jesus Work in This Death Factory?"
"The clergymen know where their paychecks come from."

The average salary of all Sandia employees, from the president
on down to the lowliest worker, was about $30,000 a year, Hosking
explained. That kind of money goes a long way in a state as poor
as New Mexico. An even distribution of the world's resources
would give everyone an annual income of about $750. No Ameri-
can can ever be that poor. "No matter how you try in this country—look, you could be a street person and you'd still have more
than $750 in assets, because of the infrastructure." He put the
hypothetical street person in a bus station and ticked off his ad-
vantages: the heat and lights, the buses themselves, the roads the
buses travel. All give even the poorest person access to more than
his worldly share of services and benefits.

Hosking was very articulate, very logical, his argument compelling enough that I put aside my momentary impulse to give
everything I owned to the poor and started wondering if even that
would be enough. "Guilt won't do any good," Hosking said, reading my mind. He gestured with his chin toward the cars roaring
out of the Wyoming Street exit of Kirtland Base. "They get irri-
tated about the signs because they feel guilty about what they do.
There's no point in that. I'm certainly not trying to promote guilt
out here and I wouldn't try to provoke guilt over the lifestyle issue,
either."

A bicyclist raced by shouting "Nuclear bombs forever!" "A significant portion of the people here say they hold to higher moral
and ethical principles," Hosking said, smiling and waving at the
parade of blurred faces and obscene gestures. "I want to point out
the inconsistencies in what they're doing. I ask only recognition
and then action. Whether Jesus ranks number two or number
twenty-five," Hosking said, "their job is number one. I don't know
how there's going to be any change as long as that's the case."

• • •

"The thing that really disturbs me about the so-called peace
movement is that they're really antipeace and don't know it," said
the quiet, gray-haired Los Alamos physicist. Clement was an ad-

ministrator in the laboratory's X-division. He had worked at Los Alamos since the late 1950s, always in a weapons design group. In 1970 he began a slow ascent up the administrative ladder, with occasional long-term assignments in Washington, D.C.

The people who went to Los Alamos immediately following the war, in the early days of thermonuclear bomb development, were the "second generation" weapons designers. Men in Clement's cohort call themselves the "second-and-a-half generation." Those who began designing bombs in the late 1960s, like Karl the French horn teacher, are the "third generation." They designed more elaborate and sophisticated weaponry. Because subsequent design work is essentially variations on themes set by the third generation, there is no fourth generation. Younger bomb designers are called just that.

The second-and-a-half and third generations have much in common. While elderly men of the first generation are most likely to feel the arms race is dangerously out of control, their protégés generally see disarmament as dangerous and are highly skeptical of arms control. Peace activists strike them as naive. "If we were to follow their folly the world would be free to engage in conventional wars," Clement explained. "And with the magnitude of firepower we have now, it would be horrendous."[2]

Manhattan Project scientists used to joke that the many intellectually brilliant and emotionally erratic Hungarian émigrés among them were actually Martians claiming to be from a mythical country because they could not disguise their heavy accents. (Included among the Martians were Leo Szilard—who coauthored the letter Einstein sent to President Roosevelt urging him to begin an atomic project and later circulated a petition opposing the use

---

[2]The first generation is not univocal, of course, but over the years many have expressed grave concerns about the arms race. After the war some founded the Federation of Atomic Scientists, now the Federation of American Scientists. The Federation publishes *The Bulletin of the Atomic Scientists*, the first public forum for information and debate on nuclear arms issues. Viktor Weisskopf and Hans Bethe have been especially vocal advocates of arms control. An interesting recent book by a Los Alamos veteran is Freeman Dyson, *Weapons and Hope* (New York: Harper & Row, 1984).

of the A-bomb against Japan—and Edward Teller.)[3] Robust and self-assured Karl, with his uncanny memory for numbers and flat and machinelike way of reporting his emotions, seemed strange enough to be an alien. Clement gave the opposite impression. He was small, quiet, and polite, with a reassuringly adult air of sincerity and responsibility. He was the family physician with a framed copy of the Hippocratic oath on the wall of his office, the lawyer you could trust, the honest statesman.

Karl didn't have much respect for peace activists. "We have these pitiful drag-tailed people who come up here in vans sometimes and hand out literature," he said. Because they were permitted to express their wacky opinions, they struck him as a monument to American freedom. But Karl did not waste a moment of his time in serious conversation with the protesters. Instead, he proclaimed, "I tell them if God hadn't loved bombs, he wouldn't have invented Japanese."

Clement made more of an effort to understand his opponents. "They don't understand me and I don't understand them. So how do we communicate?" he wondered. Once when he was in Washington he went to talk to protesters in front of the Pentagon. "Of course they expected a guy like me to have horns. Our worlds are so different, mentally. It's so interesting to talk with them." They were strong on visions of peace, he thought, but weak on realistic mechanisms to ensure it. Thus he felt reassured that the mechanisms he had helped design at Los Alamos were moral necessities.

Clement's own mental world was formed by an impoverished rural childhood in a devout Southern Baptist family. His father had once been a mathematics teacher, and when Clement finished his master's degree in physics he decided to work at Los Alamos. "Most of the world feels that I'm doing dirty work," Clement said calmly, "although I'm convinced I'm not."

---

[3] John Manley remembered the "Men from Mars" joke and appreciated its humor. See also Rhodes, *Making of the Atomic Bomb*, ch. 5, pp. 104–133. All-American junk-food fanatics—Martians who watched MTV before landing at Lawrence Livermore National Laboratory—are described in William Broad, *The Star Warriors: A Penetrating Look into the Lives of the Young Scientists Behind Our Space-Age Weaponry* (New York: Simon and Schuster, 1985).

Clement did not always find working at Los Alamos a heavenly experience. Like every other lab manager, he was burdened by paperwork and frustrated by the lab's most famous dance, the organizational shuffle.[4] But he accepted the responsibilities of administration. And Clement had had experiences Karl envied. He had witnessed the thermonuclear blast at Eniwetok and a whole series of aboveground explosions before the test ban drove the bomb underground.

"The first one I saw was high altitude and a few kilotons," Clement remembered. "It was just a flash, and I wasn't very impressed." Then he saw bigger tests—in the one-hundred-kiloton range—from a closer vantage point. "When they go off . . ." He searched for words. "They're awesome. I've never known how to describe them." Clement thought for a minute and said, "They made us wear dark glasses, of course, and long-sleeved shirts."

Long-sleeved shirts? For protection from the blast, Clement explained. He was temporarily blinded by his first vision of a very large blast. In his excitement at seeing the huge bright cloud, Clement had pulled off his sunglasses too quickly.

"On those big ones, the mushroom cloud continues to rise for a long time, and it finally runs out of power to rise upward and starts to spread over you. And that gives you quite a pause, especially when you know what's in them," Clement recalled. The sound of the blast, he explained, was less impressive. "It's just like thunder," he said. "Think of a thunderclap ten miles away. The sound is distorted."

Much of Clement's conviction of the righteousness of his work was set when he first decided to accept the job offer from Los Alamos. Clement and his wife had traveled to job interviews all over the country. "Being a devout individual myself, I prayed about it," he said. "I'd applied to Livermore. I actually *very* much wanted to go to the West Coast. It was green and lovely in April when we were interviewed."

---

[4]The charge that the federal laboratories are indeed overly attentive to management—to the point of suffering from "micromanagement"—is substantiated by an investigation performed under the auspices of the U.S. Office of Science and Tech-

They went home and waited for offers. Several divisions at Los Alamos wanted to hire him. They set deadlines for his decision. Clement waited until the last possible minute before calling Los Alamos and committing himself to a position. The next day, he got the coveted offer from Livermore. The coincidence of timing seemed a delicate form of divine intervention.

"In a sense, God said, 'Okay, go there,' " Clement said sincerely. "I do believe God directs our lives day to day." But, he added, God also gives people a lot of leeway. That's why Clement prayed for guidance, and why he felt so certain that working on nuclear weapons at Los Alamos was exactly what God intended him to do. "It's kind of neat when your Father leans over and says, 'Son, you did well,' " Clement said. "And I felt that."

•   •   •

Clement's moments under the mushroom cloud may turn Karl green with envy, but people of faith ache for the quieter experience, the feel of God's breath in their ear. Without direct evidence of God's will, it is harder to interpret moral principles and apply them to one's own life.

Jiminy Cricket's entreaty to let your conscience be your guide makes sense for children. Social psychologists interpret the still, small voice as internalized demands of our culture. It nags us with the voice of social authority, they say. From this perspective, doing what is right means conforming to social expectations you have accepted for yourself. Children are close to those lessons about good character and proper behavior. In grown-ups the interior autopilot unfortunately often steers off course. In the problematic and corrupt society of adults, social expectations may bear little resemblance to eternal standards of right and wrong. And does the conscience really guide, or does it more often just rationalize or give the illusion of free will? The human capacity for self-deception seems nearly limitless.

nology Policy, *Report of the White House Science Council: Federal Laboratory Review Panel* (Washington: U.S. Government Printing Office, 1983).

The vagaries of the conscience means that people often must search outside the heart and head for universal moral principles. God may have given each of us a bit of Himself to guide our paths, but just in case the sound of His internalized voice is muffled by self-interest, He also gave us the ability to reason to right.

"The arms race is one of the greatest curses on the human race; it is to be condemned as a danger, an act of aggression against the poor, and a folly which does not provide the security it promises." So said the National Conference of Catholic Bishops in May of 1983, in its Pastoral Letter on War and Peace, *The Challenge of Peace: God's Promise and Our Response*.[5]

Lloyd was one of the Los Alamos Catholics who strongly disputed that claim. He was in his early forties, a nuclear engineer and father of four who had been at Los Alamos for six years. Like many at Los Alamos, Lloyd studied lasers.

In 1989 researchers at the University of Utah announced that they had achieved a new type of fusion: so-called cold fusion. Shocked and skeptical, many physicists and chemists dropped whatever they were doing and rushed out to buy up the world's supply of palladium wire so they could try to replicate the Utah experimenters' "fusion in a jar." (The jury is still out on cold fusion.) The common wisdom among scientists was that only something as hot and powerful as a fission reaction could cause fusion. In a hydrogen bomb, the fusion reaction set off by a fission reaction is uncontrolled and thus destructive. But physicists also know how to control fusion reactions in a laboratory.

There are two conditions under which you can create controlled fusion reactions, Lloyd explained. Matter can be contained within a doughnut-shaped magnetic field in a machine called a TOKO-MAK. Magnetic confinement fusion is a very complicated and tricky process. Researchers at Sandia and Los Alamos researchers work on another process for producing fusion: inertial confinement fusion. A tiny bit of matter is blasted with high-intensity energy. At Sandia, they blast the targets with ions generated by two parti-

[5] National Conference of Catholic Bishops, Washington: Office of Publishing Services, no. 863, United States Catholic Conference, 1983, p. v.

cle beam fusion accelerators. At Los Alamos, they use lasers. At both labs, inertial confinement fusion has "classified applications."

These classified applications lie in the characteristics of the tiny glass bubbles that serve as the targets for the concentrated energy beams. Some of the lasers used at Los Alamos fill rooms half the size of a football field, most of the space crowded with engines that pump energy. A series of precision-ground mirrors and lenses narrows and channels this energy into an increasingly tight beam. Focusing this beam on a target that contains hydrogen isotopes like deuterium or tritium can produce, for a fraction of a second, a fusion reaction.

Researchers in the laser facility at Los Alamos told me that the weapons applications of their research are quite indirect. If the United States and the Soviet Union agreed to a comprehensive test ban treaty, both parties would be prohibited from further testing of thermonuclear weapons, period. Underground tests are the major source of scientists' understanding of the fusion process. Without testing, they can't continue to develop thermonuclear weapons—which is one reason why laboratory officials oppose a comprehensive test ban.

In congressional testimony in the mid-1980s, a Los Alamos spokesman explained the weapons laboratories' other major objection: "Given real-life pressures and fallible human nature, we might occasionally stockpile new but interesting designs or allow inadequately reasoned changes in old ones. Experience shows that serious unreliability could result." Whoops. Someone crossed out "interesting" in the testimony transcript and penciled in "immature."[6] Despite the argument that bans on testing might make the

---

[6]"Statement of C. Paul Robinson, Principle Associate Director, National Security Programs, Los Alamos National Laboratory, University of California, Before the House Committee on Armed Services Special Panel on Arms Control and Disarmament, September 18, 1985." My copy with the corrected "error" was courtesy of New Mexico Senator Jeff Bingaman's office. For a history of the Los Alamos laser programs see Keith Boyer, "The Laser Programs," pp. 80–81; and Robinson, "The Weapons Program: Overview," pp. 111–3, both in *Los Alamos Science*, vol. 4, no. 7 (Winter/Spring 1983).

In 1988 Los Alamos physicist P. Leonardo Mascheroni claimed that his research on X-ray fusion had lost its funding because laboratory officials feared it would in

arsenal less reliable, arms control specialists consider it an indispensable step in halting the arms race.

When a powerful ray of energy from a laser blasts the minuscule hydrogen-filled targets, the pressure exerted on their thin glass shells, imploding on the gases within, causes the same type of reaction that atomic fissioning produces in an H-bomb. Photographs of the basic process are mounted on the wall of one of the Los Alamos laser facilities. Greatly magnified, the tiny glass globe dimples, cracks, and then explodes, spewing out a stream of gas and glass. "It resembles a small H-bomb. It is in *fact* a small H-bomb," Lloyd said, explaining why his research was funded by the Office of Military Applications at the DOE. In essence, the classified bubbles are extremely compact versions of the Nevada Test Site. Others at Los Alamos implied that this research might also be relevant to SDI laser weapons.

I have in my possession an unclassified laser target. My guide on an informal (that is, unauthorized) tour of the laser facilities at Los Alamos casually picked it out from among several dozen lying in a cardboard carton shoved under a workbench.

The target is roughly the size of a grain of salt. It was formed at the end of a hollow glass tube that narrows to a filament no wider than a hair. For protection, the narrow tube is anchored to a bit of foam glued on the inside of the lid of a small glass vial. When you unscrew the cap, the target on its slender stalk comes out like a medicine dropper from a bottle of children's nose drops.

The label on the vial said the target was coated with a thin layer

---

fact eliminate the need for underground testing. He charged the University of California, the laboratory's operator, with violations of his academic freedom. Laboratory officials responded that his research was too much of a long shot, too expensive, and too impractical. Mascheroni's case was very messy: he also accused the university of anti-Hispanic discrimination after being charged with violating security rules and laid off. Based on what Los Alamos scientists have told me about the laser fusion program, but with no direct knowledge of Mascheroni's case, I would guess that the reason given by laboratory officials for cutting his funding is at least partially true. It seems unlikely that a truly promising route to easy laser fusion would be blocked by laboratory officials, who tend to be long-sighted, often initiating projects rather than simply responding to military requests. See Eloise Salholz et al., "An Identity Crisis at the 'Mesa of Doom,'" *Newsweek*, Oct. 31, 1988, p. 30.

of gold. Even so, the small sphere is nearly invisible. After opening the jar and lifting the target out a half-dozen times, I accidentally knocked the filament against the wall of the vial. The delicate bubble was detached from its glass tether. It lies somewhere in the bottom of the little jar, indistinguishable to the naked eye from the few specks of dust beside it, a small, fragile H-bomb.

• • •

Lloyd thought that a lot of people may not experience *any* moral dimension to life. "I haven't done any study," he said, "but I think most people just do what they know how to do."

It was one month before the presidential election, and we were drinking coffee on the patio of the Los Alamos Lab cafeteria during the midmorning lull.

Lloyd looked disgusted and shook his head. He was a conservative Republican and knew how he'd vote. Later he joked that the main difference between conservatives and liberals is whom each assumes to be free of original sin: For the conservatives, it is members of the military and business communities. For liberals, sociologically oriented politicians seem pure as driven snow.

Lloyd was devout, but he'd vacillated in his allegiance to the Catholic church. On the one hand, he missed the good old pre–Vatican II days when the rules were clear and you ate fish on Fridays. "It seemed like a nice machine to be in," he remembered. On the other hand, the church's stand on birth control struck him as ridiculous, and Mariology—the myths about the Virgin Mary—seemed even sillier. Because Lloyd revered consistency in thinking and between ideals and actions, he'd started attending his wife's conservative Protestant church. It was a way of avoiding hypocrisy, he thought, although he still considered himself Catholic.

The Catholic bishops' Pastoral Letter on War and Peace received plenty of advance publicity and helped create an atmosphere of critical reflection on the arms race. In New Mexico, some Catholics began an annual ritual designed to bring peace to Los Alamos: Every April since 1982 a group of peace pilgrims has scooped sacred soil from the floor of the Sanctuario de Chimayo, near Española. The soil is believed to heal the sick. After suitable prayers, runners carry the dirt twenty miles uphill to Los Alamos and sprinkle it on Ashley Pond. Over the next several years, Jehovah's Witnesses in Los Alamos began debating whether the prominent role of nuclear weapons research at Los Alamos was consistent with their church's doctrinal opposition to participation in secular government. Jehovah's Witnesses do not salute the flag or bear arms. Some quit their jobs at the laboratory, and one church elder predicted that most of the laboratory's Witnesses—about one-third of the sixty-member congregation—would eventually leave the lab.[7]

When Lloyd heard that people were preparing a rebuttal to a draft of the American bishops' pastoral letter, he joined fifteen Los Alamos Catholics in a year's worth of study and discussion. By the end of the year, most had drifted away. "There were a lot of cold warriors in that group," said one Los Alamos liberal who'd abandoned the project. "I got tired of arguing." Lloyd stuck with the group, took on the role of devil's advocate, and wound up as one of six principal authors of *Nuclear Weapons and Morality: A View from Los Alamos.*[8]

The cartoon cricket's rule for good behavior was simple. Real ethical reasoning is much more difficult. The Los Alamos Catholics' study group spent "approximately one thousand man-hours" on their project, producing a document of just over one hundred pages (about the same length as the bishops'), typed on a word processor and bound with a spiral plastic spine and an orange-brown construction-paper cover.

"From the very beginning," the authors wrote in their foreword, "we recognized the fact that open and frank discussions would be vital to this examination, and that such discussions would be quite unlikely in a hostile, confrontational environment."[9] Therefore, they invited only people directly involved in weapons work at Los Alamos to join their study group.

---

[7] Martha L. Man, "Witnesses Quitting Lab Over Arms Work," *Albuquerque Journal,* July 20, 1985, p. A-6.

[8] Daniel E. Carrol et al., *Nuclear Weapons and Morality: A View from Los Alamos* (Los Alamos: Immaculate Heart of Mary Catholic Church, 1983).

[9] Ibid., p. i

Both the bishops and the weapons scientists dealt with the moral justifications for war, the maintenance of large nuclear arsenals, the doctrine of deterrence, and the use of nuclear weapons. The Catholic bishops' pastoral letter relied heavily on traditional criteria for justifying military action. The Los Alamos Catholics came to very different conclusions.

The so-called just war theory, developed primarily by St. Augustine, is relatively simple. War is justified only when declared by competent authority acting with good intentions for a just cause. Claims to a "just cause" must be tempered, because there can be legitimate disagreements that do not justify killing, and no one has a monopoly on the claim to righteousness. War should be the last resort in resolving disagreements and, to prevent irrational or futile slaughter, must have a reasonable probability of success.[10]

Furthermore, the decision to wage war and the conduct of nations in battle should respect the principles of "proportionality" and "discrimination." Proportionality refers to the calculation that the harm done by a particular military strategy will be outweighed by the anticipated good results—a kind of cost-effectiveness. The benefits of military action must exceed the costs—all costs, not just military and material costs. The military action must be rationally and objectively worth it. Proportionality thus judges the means by reference to the ends, an ethical position known as "consequentialism." Utilitarianism (the idea that the best course of action produces the greatest good for the greatest number) is a form of consequentialism. This much-maligned ethical principle does not play a major role in Catholic moral thinking.

Instead, Catholic thinkers more frequently rely on "deontological" reasoning, which boils down to something like the Golden Rule. Even if acting truthfully and honestly toward people leads to unhappy results, deontological thinkers insist you must not commit moral wrongs for the sake of some seemingly greater end. Consequentialists generally allow you to throw someone overboard if everyone is otherwise doomed to die in an overcrowded lifeboat. Deontologists say no to any form of human sacrifice.

---

[10] A good introduction to the topic is Michael Walzer, *Just and Unjust Wars: A Moral Argument with Historical Illustrations* (New York: Basic Books, 1977).

Deontological standards underlie the bishops' insistence on the principle of "discrimination," which prohibits direct attacks on noncombatants and nonmilitary targets. Although it might seem obvious that only soldiers should die in war, warriors have often ignored the obvious. Of course the innocent should be spared, but what makes someone "innocent"? An innocent during wartime traditionally has been defined as someone who poses no threat to the life of a soldier.

The conduct of World War II definitively changed the world's viewpoint on that topic. The bishops outlined the moral problem:

> Mobilization of forces in modern war includes not only the military, but to a significant degree the political, economic, and social sectors. It is not always easy to determine who is directly involved in a 'war effort' or to what degree.[11]

On the ground, unarmed peasants may seem threatening, as U.S. soldiers repeatedly found in Vietnam. From the air, a city full of patriotic enemies may seem a reasonable target. Both noncombatants and nonmilitary targets now supposedly comprise a significant portion of each superpower's nuclear target lists. When the bishops investigated the United States' secret nuclear war targeting plans (the Single Integrated Operational Plan, or SIOP), they found evidence that there were at least sixty "military" targets within the city of Moscow and about forty thousand military targets for nuclear weapons in the Soviet Union as a whole.[12]

The world may accept that whole societies now go to war, but the bishops nonetheless listed categories of people who should never be considered combatants and may never be directly attacked: hospital patients, the elderly, the ill, average industrial workers, farmers, and schoolchildren. They foresaw no situation where the initiation of nuclear war could be morally justified, and expressed extreme reservations about the morality of holding societies hostage through the strategy of nuclear deterrence. Al-

---

[11] *Challenge of Peace*, p. 34.

[12] *Challenge of Peace*, p. 57, citing Solly Zuckerman, *Nuclear Illusion and Reality* (New York: Viking Press, 1982); and Thomas Powers, "Choosing a Strategy for World War III," *Atlantic*, Nov. 1982, pp. 82–110.

though deterrent strategy has been elaborated and refined over the years, it still boils down to Mutual Assured Destruction (MAD). Neither superpower dares instigate a war for fear of nuclear retaliation on its own nation.[13]

To move us away from the hostage scenario and the threat of nuclear war, the bishops called for immediate, bilateral, verifiable agreements to halt testing, production, and deployment of new nuclear weapons systems; a comprehensive test ban treaty; the strengthening of global authority to regulate international disputes; and deep cuts in superpower arsenals. Six months later, Pope John Paul II asked scientists to reconsider their involvement in "the laboratories and factories of death."[14]

The bishops' letter probably spurred the United Methodist bishops to their own denunciation of nuclear war and weaponry in early 1986. Sanford Lakoff and Herbert F. York, two longtime observers of science policy, believe that the bishops' letter and growing public and congressional support for a nuclear freeze predisposed President Reagan to seek a high-tech defense against nuclear weaponry. Archbishop Robert Sanchez of Santa Fe had a less ironic response: he joined with twelve other prelates in a campaign of nonviolent resistance against the rail shipment of nuclear warheads, the men in black joining other protesters in prayer against the white trains. For obedient Catholics, the bishops' letter and the Pope's implicit call for researchers to abandon their work on the bomb were bad news.[15]

But the Los Alamos Catholics' study group concluded that both

---

[13] For a good basic introduction to the issue, see Richard Smoke, *National Security and the Nuclear Dilemma: An Introduction to the American Experience* (Reading, MA: Addison-Wesley Publishing Co., 1984); and *Arms Control and the Arms Race: Readings from "Scientific American"*, introductions by Bruce Russet and Fred Chernoff (New York: W. H. Freeman and Co., 1985).

[14] The bishops' recommendations are summarized in *Challenge of Peace*, pp. iii–viii; Philip M. Boffey, "Scientists Urged by Pope to Say No to War Research," *New York Times*, Nov. 13, 1983, p. 1.

[15] Eric Pace, "Ban on A-Arms Urged in Study by Methodists," *New York Times*, Apr. 27, 1956, p. 1; Sanford Lakoff and Herbert F. York, "Why SDI," *Journal of Policy History*, vol. 1, no. 1 (1989): 44–79; Man, "Archbishop Sanchez Joins Nuclear-Shipment Protest," *Albuquerque Journal*, May 24, 1984.

deterrence and the arms race were morally justified responses to totalitarianism. How could two groups, with the same intentions, armed with the same holy book and the same principles, through logical reasoning reach such different answers to the same question? The Los Alamos Catholics began by dismissing the just war requirements of discrimination and proportionality.

The validity of the combatant-noncombatant distinction in modern society seemed questionable to them, since "If we all don't share in weapon building and use, we all do share in being causes of war." [16] They explained their position immediately, on the second page of *Nuclear Weapons and Morality*, using reasoning reminiscent of Chuck Hosking's argument that even the poor in America are rich from our infrastructure—our bus stations:

> It has become clear to us in our discussions that while we are very directly involved in nuclear weapons work, many others also share involvement but may not realize it. In the broadest sense all people in our country share involvement because all benefit from the protection afforded by our military defenses.
>
> People in New Mexico and Los Alamos share the benefits even more directly. A little-noted recent change in federal assistance to the Los Alamos schools and county means that now everyone in Los Alamos is financially supported by weapons funds. Because much of our county's land is U.S. government property and not taxable, the Department of Energy (DOE) contributes about six million dollars per year to the county and schools in lieu of property taxes. These monies come from DOE's Office of Military Applications (OMA), the weapons branch of DOE. Therefore, each family in Los Alamos gets about $1,000 per year in tax relief from the weapons program. [17]

And, therefore, their own children are responsible for the arms race?

In Amarillo, A. G. Mojtabai found a community filled with people who found no greater pleasure than arguing the finest details of Scriptural interpretation. [18] The Los Alamos study group showed

---

[16] Carrol et al., *Nuclear Weapons and Morality*, p. A-10.

[17] Ibid., p. ii.

[18] Mojtabai, *Blessed Assurance*.

no such appreciation of ethical subtleties. Dale Arnink, the Unitarian Minister in Los Alamos, was astounded. "Their moral analysis and development of a position is almost nonexistent," he wrote in his own twenty-page mimeographed response. He found their conclusions "shallow and naive, with little indication in context that they don't deserve the labels 'doctrinaire' and 'foregone.'" Robert Magirl, a member of the Archdiocese Peace and Justice Commission, wondered in print why the Los Alamos group "did not avail themselves of more diverse input: from women, from minorities, from those outside the establishment, and even from those inside it who hold views more in harmony with the 'conclusions' of 238 American Catholic bishops."[19]

The criticisms of *Nuclear Weapons and Morality* seemed to have little effect on its authors. "I think I decided it was okay to work on them," Lloyd remembered a year later. Defensive weapons (like those proposed for the SDI program) seemed like a better idea—"I don't like the idea of holding everybody hostage in both countries." But he confessed that the problem of thinking rationally about the morality of nuclear weapons and war could drive you crazy. The best way to deal with it, Lloyd thought, was illustrated by a friend of his who had refused a transfer to the Nevada Test Site.

"He said he didn't see anything wrong with working on them, but he didn't want to look back on his life and say, 'That's what I accomplished.' It just seems like a waste. And I personally consider weapons designing to be dead-end science.

"I would feel disappointed if I had to work on a bomb project," Lloyd said. Would he do it? "To keep my job at the lab, I'd probably work on it . . ." He hesitated just a second, then added, "Well, I *would* work on it."

Lloyd had recently refused an offer of an interesting position with the CIA. "It wasn't just the traveling," he explained, although he balked at leaving his family for long stretches on the road. "I

---

[19] Dale Arnink, "A Critique of *Nuclear Weapons and Morality: A View from Los Alamos*," mimeographed paper; Robert Magirl, *Nuclear Weapons and Morality: Another View*," *People of God*, vol. 2, no. 7 (Oct. 1984), published by the Archdiocese of Santa Fe.

think I was afraid I'd find out something really dirty and I'd have to do something about it—something really drastic, like going to Jack Anderson or something." He thought people had a social responsibility to give up their careers if they were used for immoral purposes. Saying no to the CIA kept him from running the risk of getting dirty hands. He elaborated on the theme: "If I'm developing a production line for Xylon-B gas to be used at Auschwitz and I find out about it, I should give that up."

The lunchroom was beginning to fill with people. Lloyd looked at his watch and suggested we go downstairs to ask his wife to join us. She worked in the Personnel Office, another well-educated Los Alamos woman resigned to a secretarial position at the lab for lack of better opportunities. Before we left, though, Lloyd had one other thought: Maybe God did approve of his decision to work at Los Alamos. He recalled an extended job search that he prayed would take him away from a dead-end engineering position in a gray city with bone-chilling winters. In their eagerness to escape, he and his wife had put their house on the market. For nearly a year no one showed the slightest interest in his skills or his home. One day he got a job offer from Los Alamos just as the real estate broker called with a buyer for the house. Lloyd wondered if God had put His hand on his shoulder and directed him to the Hill.

"I'd like a divine revelation," Lloyd sighed. He felt that God forgave him his sins. But often God seemed hidden, withdrawn, absent. Lloyd somewhat sheepishly admitted his latest theological proposition. After much reflection, he said, he was almost certain that people had a *right* to objective revelations from God.

•　　•　　•

"You got to have a job, security—you got to make your living somewhere," one Catholic technician at Los Alamos said, shrugging off the infighting about the application of moral principles to the nuclear weapons business. "We can't all be bishops."

Another Catholic had found the sunny side of the bishops' letter. He was an administrator at Sandia Laboratories who served on the Archdiocese's Peace and Justice Commission. "The bishops have only one place where they make a statement with no place for equivocation—they say that the intentional attack on innocent civilian populations has no moral justification." That was good, he thought, because it reminded everyone of the horrible precedent set by Axis and Allied urban bombing raids in World War II. When Archbishop Sanchez asked his opinion of *The Challenge of Peace*, he used the argument made by one dissenting bishop: The major problem for mankind is not nuclear weapons, but our proclivity for sin.

A twenty-five-year-old Santa Clara Indian technician at Los Alamos said he was Catholic. He'd been taught to believe it by his grandmother and by people in Albuquerque and Denver, where he'd been sent away to Indian schools. He did not care about the bishops. He would not work on weapons, he said, because they are destructive. He found their destructive potential inconsistent with the peacefulness of the land. He built concrete forms and did skilled manual labor for a nuclear reactor research group but even that seemed bad: "It's filthy," is all he would say. What he really cared about was his pottery, and for most of four hours he explained how his grandmother had taught him to find the clay in the riverbed near the pueblo, dig it, dry and sift it, wet and shape it into long snakes. The snakes coil into pots that he smoothes with a stone and then carves with deep designs. The pots cook in a smoky fire and turn shiny black.

• • •

The Episcopalian priest who'd said Mass outside Los Alamos Lab on the fortieth anniversary of the Hiroshima bombing sputtered with outrage when he learned that one of his brethren had worked at the laboratory since 1950. He emphatically insisted that the priest could not possibly be right with God if he worked at a weapons laboratory.

The Los Alamos priest had his own ideas. The bishops, Phillip said, "have departed from the faith once delivered to the saints. They've made a very fine attempt to defuse a bad situation, and I don't deny that the situation is very bad. But what I say is, Why didn't you say the same thing about the bombing of London? Why didn't you say the same thing about the bombing of Hamburg? Why didn't you say the same thing about the bombing of Tokyo?"

Phillip had a late vocation. He had studied the chemistry of conventional explosions for eight years at Los Alamos before he felt called to the ministry. After his ordination, Phillip kept his job researching explosives and became the associate pastor for the local Episcopalian church. But he'd recently had a falling-out with the church and resigned his post.

Part of the problem was the Episcopal bishop in Albuquerque, characterized by Phillip as "a militant antibomb man, a militant nukie-freeze guy." The bishop was forced to mitigate his rhetoric a bit when the Los Alamos diocese threatened to withhold its tithe, Phillip reported dryly. More distressing and pivotal, though, was his church's toleration of both liberation theology and charismatics. "They open themselves to a spirit," he said, explaining the problem with charismatic Christians, "and of course it's the unholy spirit that jumps right in."

If character came in colors, Phillip's would be gun-metal gray. His comment on the nukie-freeze guy was his only departure from crisply precise, grammatically correct sentences. His crew cut was ashes and lead, his eyes steel blue, his back ramrod straight even when he slouched in the padded chair in Los Alamos Lab's Oppenheimer Memorial Study Center. The sixty-three-year-old priest explained his position in what he called "a Jesuit's argument":

"I make a bomb. The fact that I make a bomb does not mean it's going to be used. I make a bomb, and it has the potential for evil as well as the potential for good. There is sin in the world. A human being cannot touch anything without the potential for sinfulness to be attached to it. The job of a Christian is to minimize the amount of evil in the world and the amount of sin. One of the ways to overcome evil actions is to fight against evil."

Someone who intends to enslave people is evil, Phillip said in an even tone, and so duty required him to do what was necessary to prevent the evil act. "That might require that I build a bomb like he has," Phillip concluded. "That's the rationale. Those are the moral principles." He fell silent.

Although we were sitting directly across a table from each other in the small, narrow room, he had turned his head to the side and kept his eyes focused on the blank wall—one hundred miles away—throughout his explanation. Suddenly he turned and looked me full in the face. "I stated it this way for an Australian interview program two months ago. I looked at the interrogator, and I said, 'Look, the United States is the good guy. This is a demonstrable fact. *We are good guys.*'"

Phillip refused to allow people to make him feel bad about his work. "My daughter used to think she should fix the world from the mess—especially the mess she thought I made." But the misguided antinuclear movement did not frighten him. "I'm not worried," Phillip said evenly. "I'll come right out and say it. I have recently toyed with the idea of looking out at the world and saying, 'Look, I'm a tired old man. I don't have many years to live. If you don't want protection from slavery, okay. Ban the bomb. It's not my world anymore.'"

• • •

It is impossible to disguise the identity of the world's only Navajo physicist. His life was documented in a *Nova* special on PBS and, he said proudly, in two films in progress, one in French, the other tentatively titled *The Legend of Fred Yazzie Begay.* Begay had found a home on the "scientific reservation" at Los Alamos. There he did basic research on laser fusion with unclassified targets.

"For many of us the question about this being a weapons lab doesn't enter," he said in the soft, hesitant speech characteristic of Navajos with English as their second language. "Sometimes our work supports weapons, with ideas. I'm still trying to live by my original convictions. I'm searching for the structure of reality. It's there. It's hidden." The weapons did not interest him. He refused to discuss them.

"There are many seminars here on weapons. I don't go to those. You look in the lab paper, you can see there are ten million of them. I stay away. But there's a little corner here . . . " He lapsed into Navajo, a language at once both guttural and sibilant. The corner was unclassified and nonmilitary. He got to use the world's fastest computers. But his work was "high risk," he added. "You get into the weapons program, you're all right. You're guaranteed to die . . ." he laughed, corrected himself, "to retire in that job." Nonweapons research was always a candidate for the budget hatchet.

Begay launched into an attack on James Watt, the former interior secretary for President Reagan. "Watt" symbolized everyone in the Bureau of Indian Affairs, past and present; everyone who had tricked Begay's people into signing unfavorable treaties; and the people who, Begay claimed, had pried open his mouth and looked at his teeth to determine the age of yet another Indian without a birth certificate. The White Man. Then Begay described his efforts to encourage young Navajos to get an education with scholarships and to serve as a responsible public representative of his people. Every so often he would slip into Navajo. Once I watched the clock over his head for three minutes as he thought out loud. He was forcing me to see what it was like to live in a world where people speak a foreign tongue. But for Begay the language difference was more than an object lesson.

"I cannot give a one-hour lecture on laser fusion in Navajo," he said, shaking his head. "There's no words. The question is, How do you do it? How do you explain it in Navajo, especially to the traditional people, the older ones? It turns out that over history the Navajo medicine men have developed a very arcane language. Within that are the concepts for laser fusion, the origins of the universe." It took Begay ten years to prepare a basic one-hour lecture on physics in his native tongue.

"The Navajo parents love it," he asserted. "They start crying. They say, 'I knew we weren't dumb and savages as James Watt says, as anthropologist says.' " The anthropologists, he explained, learn things like how to order a cheeseburger in Navajo.

Begay believes an ancient intuition lies behind the Navajo language. When he began studying advanced physics and mathematics he was shocked. "It was like flashes of light going on," he remembered. "I think, 'You've been through this stuff before.' " He felt sure that the original concepts of Western science were similar to traditional Indian spiritual beliefs. These beliefs are simple ideas, made complicated by anthropologists.

"We have a Great Spirit, he created the universe. But in order to understand it the right way . . ." More Navajo. "We talk about Mother Earth. It's our *mother*. The sky is our father. He married the two. And it created all the laws of the universe. *That's* what we call marriage. And that created all of the children. And what are the children?" he asked rhetorically.

"It's everything you would think of as the real world."

10

ENEMY LINES

.
.
.

Checkpoints and roadblocks dot the highways in the southern part of New Mexico. Traffic is delayed while officers in the forest green uniforms of the Immigration and Naturalization Service (INS) peer into each car and inquire about the birthplace and citizenship of its occupants. Reactions to the Border Patrol are mixed in the border states. Some people applaud the efforts to keep illegal immigrants on their own side of the Rio Grande. (The flat brown river is called the Rio Bravo in Mexico.) Other Americans resent having to prove their citizenship because they have a Latin accent or brown hair, dark eyes, and dark skin.

INS agents become more visible the farther south you travel. But throughout New Mexico, people pay a lot of attention to Latin American politics. One Albuquerque church was nearly destroyed by a political and moral debate centering on Latin America. The issue was sanctuary.

Saint Andrew Presbyterian Church is in the "far northeast heights" of the city, meaning it is closer to the Sandia Mountains and the Sandia Indian Reservation at the northern boundary of the city than to the Rio Grande Valley and the older downtown area. Economics and politics both tend to follow the map in Albuquerque and New Mexico. On the political and economic map of the city, the heights are economically secure and politically conservative.

Water-filled columns of translucent plastic line the south wall of the sanctuary in the unpretentious half-brick, half-adobe Saint Andrew Church. Inside, a semicircle of three hundred comfortably upholstered chairs faces the freestanding Danish modern slab of teak that serves as an altar. The bulletin board in the hallway seeks recruits for the church's softball team, Bible study, and missions in Mexico.

The sanctuary movement developed in the early 1980s when Christian missionaries returned from Central America with the word that refugees were pawns in a U.S. foreign policy game. By the mid-1980s about one hundred churches throughout the United States had formed an underground railroad to smuggle and shelter refugees from political oppression in Latin American countries. These latter-day Harriet Tubmans deliberately avoid secrecy. The churches formally announce their sanctuary commitment. The fugitives they sponsor may cover their faces in public meetings, but they do speak out. A sanctuary church may do nothing more than offer philosophical and financial support to victims of oppressive regimes, but it may also provide volunteers to physically transport and shelter Latin Americans seeking political asylum. This is a serious moral responsibility. It is also illegal.

The status of refugees depends on which country they flee. The Reagan and Bush administrations have opposed the Sandinista regime in Nicaragua and supported the governments of Guatamala and El Salvador. Although the latter two have been labeled among the world's worst human rights offenders, only Nicaraguan refugees have a real chance of gaining political asylum. Accepting the claims of Guatamalans and Salvadorans to political asylum would imply that two governments friendly to the United States are violent and repressive regimes.

By 1984 a few activists in the sanctuary movement had been

indicted on multiple felony charges for violations of federal law, and other indictments were on the horizon. Saint Andrew Presbyterian would have been the first sanctuary church in New Mexico. Sponsors of the proposal hoped to have a formal declaration of sanctuary on July 4, 1985. Their national organization, the General Assembly of the Presbyterian Church, had already endorsed the sanctuary movement.

The church's Christian Social Response Committee spent six months organizing educational programs and discussions on the sanctuary proposal. The church elders voted two to one in favor of declaring their church a sanctuary. Normally that would be enough to establish church policy.

But Saint Andrew Church had a special problem. About 10 percent of its membership worked at Sandia Laboratories or Kirtland Air Force Base. When the Saint Andrew congregation began discussing sanctuary, lab workers promptly informed their supervisors. "Keeping the supervisor informed is a matter of personal protection," noted William Miller, a church elder and chair of the Christian Social Response Committee. Some Sandia Laboratory supervisors responded by threatening to revoke the clearances of Saint Andrew churchgoers if the church became a sanctuary. No clearance, no job.

Because the issue was so very divisive and controversial, the elders agreed to put it before the general membership. The congregation split three ways, with one-third voting for sanctuary, one-third voting against, and the rest asking for an indefinite delay on the decision. The compromisers won, but the painful debate nearly destroyed the church, and their dismayed pastor resigned shortly thereafter.

The security clearance issue raised thorny questions about the separation of church and state. Could a Presbyterian be forced to leave his church to keep a government security clearance? Saint Andrew Church discovered that a Sandia supervisor can in fact revoke an underling's clearance, although the employee has a right to appeal the decision. The church had surveyed its membership before bringing the matter to a vote. The questionnaires revealed that regardless of their individual situations, some people had been spooked by the threat of government surveillance. Wrote Miller, "[O]ur members who actually *held* a clearance were no more likely than others to oppose sanctuary, to doubt that they could be protected, or to indicate that they would not be personally involved. These members were not 'voting their jobs.'"

Sanctuary supporters suggested that lab employees might send a letter to their supervisors explaining the church's position and promising that they would not engage in any illegal sanctuary activities. This suggestion was less helpful than it seemed. As Miller put it, "To send a letter stating, 'I will not transport, shelter, feed, clothe, or otherwise knowingly assist any undocumented Central American' proved a jarring contrast with the gospel."[1]

The Albuquerque Society of Friends (Quakers) became the state's first sanctuary church. In 1985 the Santa Fe City Council approved a resolution declaring the city a sanctuary for Salvadoran and Guatamalan refugees. The following year, New Mexico Governor Toney Anaya, in an unprecedented and highly controversial move, designated the entire state a sanctuary. Anaya's sanctuary proclamation was quickly rescinded by his Republican successor.

Nonetheless, Anaya's move probably saved two New Mexico sanctuary defendants from a term in federal prison. The Reverend Glenn Remer-Thamert had smuggled two pregnant Salvadoran women into the United States. Remer-Thamert had been to Salvadoran orphanages and did not like what he saw. He blamed the terrible poverty in that war-torn country on the political conditions, and had adopted a Salvadoran child himself. He promised to help find adoptive parents for the two women's babies.

When they crossed the border, Remer-Thamert and the two Salvadoran women were accompanied by Demetria Martinez, a journalist who covered religious news for the *Albuquerque Jour-*

---

[1] William Miller, "Sanctuary and Security Clearances," a paper prepared for the St. Andrew Presbyterian Church congregation, 1985. I am indebted to Miller and to Joyce Mitchell, another St. Andrew's elder, for their help in understanding the issues facing the church. The church did go on to pursue a variety of legal means of affecting U.S. Latin American policy and supporting political refugees. They are described in Miller, "What Has Become of Sanctuary at St. Andrew?", a paper prepared for the St. Andrew Presbyterian congregation, no date.

nal. Some of Remer-Thamert's expenses had been reimbursed by a private adoption agency. Federal prosecutors claimed that about $2,000 of those expenses were unsubstantiated. In 1988, a few days before Christmas, the minister was charged with baby selling. The journalist was named as his accomplice.

Both were acquitted. After the trial, jurors said that two things had swayed them. Martinez, they thought, was protected by the First Amendment guarantee of freedom of the press. Both she and Remer-Thamert had another compelling argument, though. The defendants believed the governor's sanctuary proclamation had also protected *them.*

"I want people to know that we're real people up here and not a bunch of diabolical robots," Elton said on the phone, explaining why he had called from Los Alamos to volunteer to talk to me. I went to his home in the early evening. Since his teenagers were busy with their homework and his wife had just put their baby to sleep, he suggested we might find more privacy downstairs.

The floor of Elton's basement was littered with hand tools and scrap lumber. As he led me around a sawhorse and grabbed two metal folding chairs from against a wall behind a rat's nest of heavy-duty extension cords, the forty-three-year-old technician explained that he was halfway through an elaborate renovation. His home was built on a hill, with the back of the house facing into a wood visible through newly installed windows in what would eventually be a rec room.

"If I don't have a screwdriver or wrench in my hand, I'm not happy." Elton was fit and energetic, talking and moving like a man in his mid-thirties. He straddled his chair, in my line of sight but facing away at a ninety-degree angle, and stared out at the trees as he spoke.

The son of an engineer, Elton grew up in a wealthy community where everybody graduated from high school and automatically went to college. He went along with the crowd but never really settled into college life. He studied mathematics and engineering, spent two years in the army, worked as a machinist, studied some more, became a technician at an electronics firm in Arizona, and eventually realized that he would never finish his bachelor's degree. He headed toward Albuquerque to look for a more interesting job. There he met someone who urged him to apply at Los Alamos.

"I was really impressed as I drove up the hill." Elton remembered the soft look and sharp scent of the dark pines at the top of the mesa. "It looked like a dream—the town, the mountains, the facility." Los Alamos National Laboratory offered him a job on the spot.

For a while Elton was a "rent-a-tech" in a support group that provided short-term expertise for projects all over the laboratory. One of the groups he consulted for was so status-sensitive that the mailboxes in its secretary's office were color coded, blue for staff, yellow for the lowly techs. Elton soon found himself a better position in the lab, where he could work independently testing small electronic devices designed by an engineer at Sandia.

The professional staff at Los Alamos may treat the technicians as highly skilled servants, but Elton considered himself a professional and was appalled by the unionization of technicians at other national laboratories like Argonne and Fermi. "We feel we have a handle on our own destiny," he explained. "We don't need anyone to do it for us." Besides, he said, the really good scientists, the "highfalutin" experimentalists, did not pay attention to academic credentials. All they cared about was intelligence and skill.

"I had a visit from a guy I held in high esteem for years and years," he reported proudly. Elton's visitor was a senior professor at the Massachusetts Institute of Technology. "He came to *my* lab today. He came and *listened* to me." Elton was high on the honor. Everything was coming up roses for him. Los Alamos was a great place to live, he said, with an efficient municipal government. The only oddity was his neighbors' regular churchgoing. It was the most religious non-Mormon town he had ever seen. He grinned and hypothesized about the cause: "We're closer to God. It's 7,300 feet up here." Then he turned serious. Maybe the altitude really *did* make a difference. "People here seem more human. They seem more real up here."

Elton did no classified research. But was it weapons related? He wrestled with the question. Nothing he had ever done at Los

Alamos was directly funded by a weapons program or directly connected to nuclear bombs. But everything he worked on for the laboratory had weapons applications. Was he doing weapons work? He tried to reason it out but the truth eluded him. Hemming and hawing, Elton finally compromised, saying that his efforts should be considered fifty-fifty.

Elton's struggle to define how much of his work was weapons related was a familiar spectacle. An older Sandia scientist had warned me to expect problems with the question. "Do you know what Sandia does?" he had asked. He predicted that a twenty-seven-year-old recently hired Ph.D. would tell me that Sandia did terrific exploratory research on important scientific and engineering problems. "And he would say that he doesn't have much to do with weapons—he's doing *research*." The older man thought that was a mistake or worse. The truth was harsh and simple. "We weaponize," he had said. "We make weapons out of physics packages or nuclear components. We are an ordinance engineering laboratory."

In fact, only two groups of people at Sandia and Los Alamos had straightforward estimates of what percentage of their work was weapons work. Those in peaceful energy research groups (like solar and nuclear power plant research) almost all said, without hesitation, that none of their work was on weapons. Those in components and weapons design groups invariably reported that between 95 and 100 percent of their time was spent on weapons work. But for people in other areas at the laboratories, estimates about their personal proximity to nuclear weapons were based on all sorts of different criteria, including some apparently not measurable.

For example, three people in the same small group at Sandia, working on the same mostly unclassified projects, had widely differing estimates of the proportion of their work related to nuclear weapons. The group did shock-wave tests. With an eighty-foot-long impact gun they created a shock wave and measured its effects on various materials and instruments.

The group's technician thought that three-quarters of their time was spent on weapons work. The group happened to include the apocryphal twenty-seven-year-old energetic and enthusiastic recently hired Ph.D. Sure enough, he guessed that about 5 percent of his work was connected to the bomb. Sometime in the next year or two he would be expected to provide someone with the details on how his research helped the laboratory fulfill its prime mission. But an older staff member, nearing the end of his long career at the lab, insisted it was wrong to ask him to put his work into a little box with a label on it. Since Sandia was paying the bills, he supposed someone might say all his work was related to weapons. He maintained that it was all basic research.

Some people figured that weapons work was anything paid for out of a DOE weapons project fund. Others identified only classified work as weapons work. Others, like Elton, tried to foresee the path of the serpent that twisted and looped through the scientific community and the military-industrial complex, connecting their own research to some eventual application to national defense, sometime in the future, in someone else's office or laboratory.

A Sandia engineer who designed computer simulations of weapons components worried the question like a puppy does an old shoe. He decided, finally, "It *is* weapons work, but I can make other people feel better about it by saying the vast majority of the components I work on are safing mechanisms. They make the weapons safer." Who those other people were he did not say. Another man never laid eyes on anything remotely resembling a bomb or its parts. But because his job was indispensable for the operation of the Los Alamos Laboratory—he was a building maintenance supervisor—and because the purpose of the laboratory was to design nuclear bombs, he labeled himself a weapons worker.

• • •

It had grown too dark to see in Elton's basement. He flicked on a light and sat down again, folding his arms across the back of his chair, still facing the windows, now black, at the back of the room. His profile was highlighted by the bare bulb hanging from a half-assembled ceiling fixture. "Everybody I know in the science business would like to be a Dr. Salk," he said. "Or an Einstein. Those are the heroes." Anyone with education and talent in science had a moral obligation to help improve the human condition, he thought. And scientists could be more useful than lawyers or most other professionals. Lawyers especially were too easily lured by the smell of the hip-pocket green. Money meant less to those inflamed by the search for truth.

But then Elton had to explain why so many people with scientific expertise do not seem to do much for the rest of humanity. He had a ready answer. Technology is neutral, he said. Every invention has the potential for multiple applications. The moral value comes from its use. Consider a hammer. You can build a church with it. You can hit somebody over the head with it.

"I *hate* weapons. I think most of the people I know, involved even in warhead design work, feel the same way. I don't like the philosophy we hear bandied about that you have to be strong and tough and always ready to go to war." On the other hand, Elton admitted, he had become an opportunist. Weapons money was the healthiest research funding around, the scientific equivalent of a corporate law practice. "I would consult on a warhead development project, not because I wanted to succeed so much as for what I can learn from it," he declared. "I don't see a nuclear weapon as any more diabolical than any other bomb. And," he added smoothly, "if we don't have it, Qaddafi might."

Qaddafi might anyway. Elton seemed to reconfigure his mind with each thought. He realized that he would not be willing to work in a weapons *manufacturing* plant, for example. And chemical weapons were particularly offensive, perhaps even diabolical.

If technology was neutral, then what distinguished nuclear weapons from chemical and biological ones, morally? Elton was not the only weapons lab employee who drew the line at chemical and biological warfare. Many people expressed a deep-seated revulsion for poison research. No one could explain why a puff of toxic gas was morally worse than the hot blast of a thermonuclear bomb and its invisible radioactive fallout. The *B* and *C* of the military alphabet (NBC: nuclear, biological, chemical) just *seemed* more evil.

Elton tried for a rational answer to the question. He started by considering the amount of pain various weapons produce. He thought we should compare how many people each type of weapon might kill or maim. He thought "diabolical" weapons had a greater chance of killing the innocent. He leaned forward and started to rock back and forth on the front two legs of his chair. When Elton first decided to go to Los Alamos his brother had written him, urging him to reconsider. His brother thought Los Alamos was "100 percent diabolical," until Elton wrote back that he did not work on weapons and explained the hammer analogy.

• • •

In 1963 Stanley Milgram began a series of experiments in which an "authority," played by the researcher himself, ordered volunteer research subjects to administer electroshocks to other people as part of an experiment on learning. Like many experiments in social psychology, Milgram's was based on deception. The shocks were faked, the "subjects" were stooges in league with Milgram, and the student volunteers were dupes.

In Milgram's laboratory, a nervous man might begin to worry that he was doing serious damage to the subject of the "learning experiment" in the room next door. He could hear the victim pounding on the wall or calling for help. Milgram would say, "It is absolutely essential that you continue." With such relatively mild commands he could convince about two-thirds of his naive subjects to deliver the maximum 450-volt shock to the "suffering" stooge. On the "control panel" the levels of shock were labeled from "Slight Shock" through "Intense Shock" to "Danger: Severe Shock." The 450-volt maximum was marked "XXX." The naive subjects were ordinary college students and ordinary members of the community, some of whom were paid four dollars plus carfare for the hour they spent helping Milgram with his experiment.[2]

---

[2] Stanley Milgram, "Behavioral Study of Obedience," *Journal of Abnormal and Social Psychology*, no. 67 (1963): 371–8. In the original experiment, Milgram gave his instructions directly to the subjects, who could hear protests from the stooge in the next room. When there was no feedback at all from the stooge, 100 percent of

Since the end of World War II researchers in social science had been preoccupied with the obvious question of why it was so *easy* for the Nazis to inflict their cruelties. Some social scientists came up with answers specific to the German and Austrian experience. Milgram's studies disrupted the social scientific consensus that it probably couldn't happen here.[3]

Many occupations are associated with moral stereotypes. Our popular culture may make cruel jokes about doctors, lawyers, ministers, judges, and teachers, but still these are treated as noble professions. The work of secretaries, construction workers, and retail clerks seems intrinsically neither good nor bad. Prison guards, dogcatchers, and foreign spies are easily cast as malevolent. Scientists and engineers catch their share of the mud slung toward the top of the pedestal.

"I just want to get out that I'm a human being and I don't like weapons any more than anyone else," Elton said, knowing that people judge character by superficial criteria, like what others do for a living. (He had apparently forgotten his indictment of the lawyers.) Elton thought that slighted the human qualities of his coworkers on the Hill. It was insulting and unfair. No one had ever actually called Elton a diabolical robot, but he figured that was the unspoken verdict. By demonstrating that he was a regular kind of guy, he could break the stereotype.

The bombheads thought they were doing the right thing. What about those who had doubts? Where would they draw the line? Many were uncertain. A few said they would refuse any weapons-related work but hoped their jobs would never depend on it. Some

---

the subjects administered the maximum shock. When left to their own devices, without orders from Milgram, only 2.5 percent of the subjects administered the maximum shock. For a description of Milgram's modifications to his original research design, see *Obedience to Authority* (New York: Harper and Row, 1974). A readable discussion of his findings is in Roger Brown, *Social Psychology*, 2nd ed. (New York: Free Press, 1986): 1–42.

[3] Social scientists have disagreed about the specific causes of public support for Nazism but generally agree that those causes were peculiar to the German and Austrian experiences. See, for example, Theodor Adorno et al., *The Authoritarian Personality* (New York: Harper, 1950); and Theodore Abel, *Why Hitler Came into Power* (New York: Prentice-Hall, 1937).

were willing to consult on weapons projects in a limited way.
Those limits were set by their attention spans. Since weapons
work was reputedly dull, scientists and engineers with more cosmic questions before them did not want to be stuck redesigning
the toothbrush. Others said they would work on the parts of a
bomb but did not want to calculate its ability to kill people. That
job seemed too gruesome, even though, as one man put it, no
blood came out of the computer. Some wondered how anyone
could stand to draw the circles around the cities on the military
maps. Others felt there was no sensible moral line.

Workers at the weapons laboratories gave two explanations for
their trouble discriminating between weapons and nonweapons
work. The effects of scientific and technological developments can
never be foreseen, some said, because cause and effect in human history are unclear. DDT and deterrence were the favored
examples. The megatons may be designed for shooting and dropping but may also be our only savior against large-scale conventional wars.

While intentions make a difference, people said, science is less
predictable than nature. Good intentions can have evil consequences. The guy you hit on the head with a hammer might be a
bad man. The church you build might shelter Nazis. The research
on fusion may give us a cheap, clean source of energy. It may also
produce better bombs. Doing weapons work may seem evil but
serve the greater good, just as "pure" basic research in universities
may be the basis for the weapons of the future.

People also complained that moral responsibility in complicated
social systems is almost never very clear-cut. You might easily ad-
mit, "Yes, father, I did it," when the question is, "Who chopped
down the cherry tree?" But if you are one of the people who design
neutron generators at Sandia National Laboratories, should you
say, "Yes, I am responsible for the bomb?" Are you really more
responsible than the president, the Congress, the ordinary citizens who elect them? "The guy who delivers the mail to the lab is
also responsible," said one Sandia engineer.

A common warning given to aspiring writers is to avoid the stylistic
tendency known as "vanishing agency." This is the impression

given when writing is done with too great a reliance on the passive voice, so that the doer, the agent, seems not to exist. The actions appear to do themselves, as in these sentences, where impressions are given and writing done by nobody.

Bureaucracies, notorious for producing memos written in the passive voice, are organizations for the anonymous. Agency there becomes a matter of job title. It is important only that each person in a bureaucracy respect the duties and limits of his or her authority as defined by the organizational chart. Clients, identified by their case numbers, are equally anonymous. In the tangled structure of society's organizational chart, the proliferation of interdependent roles can make decisions about responsibility seem either trivial—simply a matter of the job description—or endless. If society is a seamless web, then every strand is equally important or equally inconsequential, every person equally anonymous.

Just as few people at the lab have direct and incontrovertible responsibility for nuclear weapons, few have the option of complete "deniability." For most people, therefore, saying "I make bombs" is a self-identification. The line it draws is part of a self-portrait. When the world seems determined to malign the bomb and eager to condemn those who make it, admitting to weapons work requires some daring. It may also be a challenge, since scientists and engineers might just as plausibly and much more easily mumble something platitudinous about the search for truth and high technology. Why embrace the stigma?

Admitting to doing weapons work can be a showstopper. Perfectly lovely dinner parties fell to pieces when someone at the table took umbrage and drew analogies between weapons researchers and Adolf Eichmann, the Nazi officer who made the trains run on time. That response was so predictable that one young Sandia engineer, hoping to liven up a particularly dismal cocktail hour, bellowed "I make bombs" when someone casually asked how he made his living. He was baffled when the announcement did not incite controversy. Everyone ignored him and continued their boring chatter. He went home early.

One night I talked with an engineer who had worked at both Los Alamos and Sandia since the end of World War II. He was pleased with his career. "I'm very patriotic and I'm proud," he said over and over. Although things had changed a lot since the old days, he had never had a moment's doubt about his work. No matter which way the winds of public opinion might blow, he knew he was doing the right thing by helping to keep America strong.

We walked together to his car in the dark parking lot of the restaurant in which we had met. We shook hands and I started to walk away. But he had one more story: He was waiting to get on a plane. A bored and restless little boy was running around the airport making a nuisance of himself. They struck up a conversation and the kid asked, "What do you do, mister?" The Sandia man replied, "I make bombs." The child shrieked in fear and ran off.

Waving his hands in the air in mock terror, the engineer mimed the scream. He was still laughing when he climbed into his car.

• • •

"**I** am concerned about the Nuclear Freeze movement, more concerned than I would be were it not coming from a central segment of our society," wrote George Dacey in a *Sandia Lab News* essay in 1983. Dacey was the president of Sandia National Laboratories until he retired in 1986. His statement was reprinted on the editorial page of Albuquerque's afternoon newspaper.[4]

> In the earlier "peacenik" movements, we were dealing with people who were on the fringes of society. Today, however, the debate features Catholic bishops, members of the American Bar Association, American Medical Association—people whose views cannot be dismissed as not representative. That is one reason, I think, that it is an important movement. But I regard it as dangerous for this reason—these people don't seem to be any better informed than people were in the former movements.

Dacey doubtless meant to lend weight to what he saw as a more authoritative consensus among responsible and well-informed people. Some Sandia workers were infuriated. They found in Dacey's statement the suggestion that those who support arms control

---

[4]George Dacey, "A Kind of Crossroads," *Sandia Lab News* (Apr. 1, 1983), reprinted in *Albuquerque Tribune*, Apr. 13, 1983.

or a nuclear freeze are fools or communist dupes. A sixty-year-old Sandia engineer who had been involved in aerodynamics research on "damn near every weapon in the Atomic Museum" thought it foreshadowed a return to the repression and paranoia of the McCarthy era.

"I'm very socialistic in my thinking," he said, stressing the last syllable carefully. He was afraid that talking about this might get him into trouble with the lab's security force, but he just had to say it: socialist philosophy seemed conservative to him. "It's a very socially responsible philosophy," he explained. "It's not a 'Do your own thing' type of philosophy." He was thoroughly fed up with the American obsession with the survival of the fittest and capitalist free enterprise. And he remembered exactly the moment when he first realized that his mind was closed and decided to open it. He was in his early thirties and had been sent by Sandia on a consulting trip to the West Coast.

"I was browsing through a bookstore one afternoon and I came across a book that just stood out there," he recalled. "*Das Kapital*. By Karl Marx." This was about the time of the McCarthy witch-hunts.

"I reached up to get that book off the shelf, and my hand just stopped. I kind of hesitated. And I thought to myself, 'Why did I do that? Why did I hesitate?'" He reached up, heart pounding, grabbed the book, and laid it right down on the counter beside the cash register.

Thirty years later, a nuclear freeze seemed like a good idea. He donated money to the Union of Concerned Scientists, too, he said. But he felt a certain anxiety about that, because donations and affiliations must be reported on the lab's annual security questionnaire. He was a liberal Democrat and was tired of being intimidated by his narrow-minded flag-waving colleagues at the laboratory. He could hardly believe they were allowing me to interview people about their political views. He could hardly believe he had called me to volunteer. Perhaps, he thought, he felt freer because he was about to retire.

• • •

One morning during the early days of my research I was drinking coffee and going over my notes from an earlier interview in the Los Alamos Laboratory cafeteria. Absorbed by my work, I was oblivious to my surroundings and was startled when someone asked if he and his friends could share my table. Morning had turned to noon and the lunchroom was packed.

Seven people sat down and started to joke with each other. It was five days before the 1984 presidential election. "The Nuclear Freeze people called me up last night, urging me to vote for Mondale," said the thin young man beside me. I started paying attention.

"And you told them?" someone prompted from across the table.

"I told them I already had, on an absentee ballot. What I didn't tell them was I was a nuclear weapons designer."

"Yeah," someone chimed in, "one bomb designer can ruin the Freeze people's whole day. We could have a group, 'Nuclear Weapons Designers for Mondale.'"

The jokes continued. Apparently everyone at the table was in X-division. The guy beside me seemed the token liberal Democrat. I sneaked a look at his identification badge as he chided the pudgy man across the table for eating onion rings—bad for the heart and the waistline, he said, and besides, they were a Democratic vegetable. Only the catsup was properly Republican. The entire party hooted and moaned.[5]

The comedians from X-division obviously assumed that I too worked for the laboratory. Allowing them to think I was one of them in order to eavesdrop seemed no minor deceit, and I felt obliged to introduce myself. I told them I was a political scientist from the University of New Mexico. "Consulting?" someone suggested. No, I explained, interviewing scientists and engineers to find out how they felt about working at the laboratory. Dead silence. The mood of the group was poisoned. Weeks later I interviewed the Mondale supporter. Several of his coworkers, he said,

---

[5]Catsup joined Teflon as a homespun symbol of the Reagan administration when the federally supported school lunch program defined catsup as a vegetable. Doubting its nutritional value, congressional Democrats and the general public raised a fuss.

were still angry that I had witnessed their backstage clowning. They realized that they had sat down at my table. Had they bothered to look, they could have seen I was not wearing a badge. But they were upset that I was not the person they assumed me to be. They felt infiltrated.

The weapons laboratories have become more open since the war years. During the second World War political dissent was suppressed at least once, when Leo Szilard circulated a petition among scientists at the University of Chicago Metallurgical Laboratory. Only if the terms of surrender were first made clear and the Japanese refused should the use of the bomb on a city be considered, the petition said. Szilard got sixty-eight signatures on his petition. He sent a copy to Oppenheimer at Los Alamos, but apparently Oppenheimer decided not to circulate it. Meanwhile, to prevent it from reaching the president, army officials classified the document and locked it in a safe.[6]

Also long gone are the days when J. Robert Oppenheimer was made an outsider, confined to the wrong side of the "brick wall" of secrecy. Oppenheimer had traveled in left-wing circles before the war. After fascism was defeated, communism became the nation's number-one ideological enemy. In 1954 the founder of Los Alamos was deemed untrustworthy by the government's Gray Commission and lost his access to atomic secrets.

Edward Teller was the most persuasive witness to cast doubt on Oppenheimer's judgment. By all accounts, Teller has a powerful personality. People who know him said that he is like a "force of nature," or a "demiurge." One Los Alamos thermonuclear weapons designer compared a conversation with Teller to a meeting with Moses.

Teller has often been accused of self-serving inconsistencies in his public statements, and I had hoped he would respond to the

---

[6]Szilard fought to have the petition declassified after the war. It was returned to him in 1957. For the text of Szilard's petition and other relevant documents see Weart and Gertrud Weiss Szilard, eds., *Leo Szilard: His Version of the Facts, Volume II: Selected Recollections and Correspondence* (Cambridge, MA: MIT Press, 1978), esp. pp. 186–8, 209–22. Rhodes notes, "The bombs were authorized not because the Japanese refused to surrender but because they refused to surrender unconditionally." (*The Making of the Atomic Bomb*, p. 698).

criticisms from some within the labs. He agreed to an interview, but the day before we were to meet he called me and imposed two conditions: I must tell him the names of those I had interviewed, and I must agree with his position on SDI.

The first condition he imposed because he felt my research, which I had described in a letter to him, might be very controversial. He wanted to know what company he would be keeping. And I must concur with his opinion of SDI, he explained, because he thought the program essential for the future security of the United States and did not want to waste precious time before the election with a skeptic.

My promise of confidentiality to the people I interviewed made the first demand impossible to meet. I also told Teller I was unsure of the potential costs and benefits of SDI. We never did meet. The military officer who had helped arrange the aborted interview later told me that Teller had claimed he'd canceled our appointment because I refused to provide a written list of questions in advance.[7]

The old-timers at Los Alamos and Sandia were still outraged by Teller's indictment of Oppenheimer. When Teller's name came up most of them thought of that incident. One Manhattan Project scientist thought Teller had been motivated by a carefully hoarded stockpile of old grudges. Oppenheimer had not always been tolerant of Teller's Martian caprices and had not appointed him to head the secret laboratory's most vital and prestigious group, the T-division.

"The main thing was that Oppenheimer interfered with the proper exercise of Edward's ego," Teller's former colleague said. "Oppenheimer didn't give Edward's ideas the prominence Edward thought they deserved." He was referring to Oppenheimer's cautious attitude toward Teller's pet project, the "Super" thermonu-

---

[7]Teller did, however, have his research assistant send me articles he had authored on the relationship between scientists and government. Included in the packet was an essay titled "The Noble Lie," from *The Reluctant Revolutionary* (Columbia: University of Missouri Press, 1964), pp. 3–24. There Teller took Plato to task for allowing government, in *The Republic*, to deceive its citizens for the supposed sake of the common good. For an account of Teller's often difficult relations with journalists see Norman Moss, "Sunday with Edward Teller," *The Listener*, June 13, 1985, pp. 13–14.

clear bomb, which Teller had pursued during the Manhattan Project while his colleagues devoted themselves to the atomic bomb. As head of the scientific General Advisory Committee to the Atomic Energy Commission after the war, Oppenheimer added insult to injury by opposing the accelerated development of the H-bomb. With the aid of damning testimony from Teller, the Gray Commission interpreted Oppenheimer's call for temperance in thermonuclear weapons development as a possible sign of divided loyalty.[8]

A twenty-five-year veteran of Sandia remembered when the lab used to hand out literature that advised weapons scientists to "Watch Your Liberal Friends." We had met for lunch at his favorite Chinese restaurant. "I think I'm as liberal as they get," Marty confided, piling egg rolls and sweet-and-sour pork from the buffet onto his plate. "I'm a socialist at heart."

Because he had acted on his political principles, Marty was regularly harassed by clearance officers at the laboratory. He said he loved to talk about it now that things had loosened up. Later Marty worried that maybe he had said too much. Would I be careful to disguise his identity?

Marty had worked on military engineering projects during World War II and saw no reason not to continue in defense work. He first discovered that his patriotism was in doubt in the early 1950s while working for a military research laboratory. One day he mentioned to his supervisor that he planned to run for the chairmanship of his Democratic party precinct. His boss begged Marty not to do it. The military security men had been making noises about the young physicist who seemed too sympathetic to controversial causes. If he became more visible in Democratic party politics both of their positions might be jeopardized. Marty thought it over and decided to keep his job.

Things got worse in the 1960s. Marty supported the civil rights

---

[8]The testimony transcript of the Gray Commission hearings in 1954 are in the United States Atomic Energy Commission, *In the Matter of J. Robert Oppenheimer* (Cambridge, MA: MIT Press, 1971). A different perspective on Oppenheimer's leftist associations is told by Haakon Chevalier, *Oppenheimer: The Story of a Friendship* (New York: Pocket Books, 1966; orig. published by George Braziller, 1965).

movement and opposed the Vietnam War. He was put under surveillance. He was once accused of sexual indiscretions, to his horror. He was just too liberal and too outspoken.

The security "interrogations," as Marty called them, eventually became a familiar routine. He even developed a small sense of humor about the situation. The security men shook with frustration and anger when Marty waxed rhapsodic about how wonderful it was to live in a country where an individual had the right to invoke his Fifth Amendment protection against self-incrimination.

After twenty-five years at Sandia, military applications had surfaced for even Marty's most farfetched and fanciful unclassified research. So how did he feel about having a political life verging on the revolutionary and a professional career devoted to weapons work?

"The fact that you're making the napalm I don't think should keep you from being against its use," Marty replied. He slouched back in his chair in the now-empty restaurant. "The fact that you're involved in making the A-bomb shouldn't keep you from being strongly involved in opposing its use. I wouldn't be comfortable if everybody who was against war—which the military would have you believe includes everybody in the army—would quit and go into agriculture." The tax-paying potato farmer in Idaho also played a supporting role in the system, he added. There is no escape.

Reluctant to leave defense work to the hawks, Marty simply drew a thick black line between his politics and his livelihood. It kept the two separated as long as the security men stayed off his back. Marty's heart was in his political work. He talked in the first person about politics. ("I voted for Norman Thomas, the socialist candidate for president.") Marty's professional work ("what you're creating") he tried to do well, but he described it in the second person. He wasn't sure if he'd actually had any impact on decisions by his colleagues and bosses at the laboratories. But he knew he was an irritant.

Officially, it is not the laboratories' business to prescribe or proscribe any political position. But the presidents of the two laboratories periodically analyze the impact of current events on the future of the labs in their in-house newsletters, communicating more than their mere personal opinions.

To hear them talk, the liberals at the weapons laboratories, activist or not, were besieged. They complained of diffuse and low-level persecution. Most who felt politically alienated or suppressed blamed it on the small-mindedness of their colleagues. The pressures were subtle but persistent. The political cartoons they posted on their doors would be torn down the next morning. They were the butt of practical jokes.

"They can buy my talent, and I'll sell it to them," one Sandia scientist said. "But they can't buy my soul." He was incensed for a small reason: He was supposed to list his most innocent hobbies on the annual security questionnaire. And so he did. "I'm a little cautious about joining organizations because of the security thing," said another, younger Sandia researcher. "It can cause some awkward questions. I feel it's better to avoid those questions than to make an example of yourself." The sacrifice was worth it, he thought. He liked his salary, his supervisor, and the projects he worked on. He was an adventurer on a fantastic frontier.

11

SHELTERS

•
•
•

"**D**avy Crockett" is about as big as a medium-sized dog or a pony keg of beer. The first time I saw the tiny nuclear weapon in the Atomic Museum in Albuquerque, its destructive radius was illustrated by a circle drawn on a map of the University of New Mexico campus. Sometime in the mid-1980s the map disappeared. When I asked the museum docent why it was taken away, he looked puzzled and said, "What map?" Someone must have decided that demonstrating the effects of a small atomic bomb on the local college students was not good public relations.

Americans were consumed with nuclear war anxiety in the early 1980s. The trend probably began with the highly publicized congressional debates over the "clean" neutron bomb, the weapon one Sandia physicist had likened to a rifle because it would kill living things while preserving the physical infrastructure. The largest public demonstration ever seen in the United States was in 1982, when 750,000 people gathered in New York's Central Park to protest the arms race. Jonathan Schell's book *The Fate of the Earth* described the land after a nuclear war as "A Republic of Insects and Grass" and became a bestseller. The Nuclear Freeze movement gained momentum.

The Reagan administration proposed a $4.3 billion seven-year civil defense initiative to Congress. Instead of allaying public fears, the civil defense program—along with suggestions from top administration officials that a limited nuclear war could be fought and *won*—increased people's sense of imminent disaster. The doomsday machine had emerged from the shadows.[1]

Perhaps the nightmare-plagued public would feel safer under some form of heavenly protection. A child's vision of a beneficent rubber rainbow was used in a television commercial to convince the public that the Strategic Defense Initiative would shield their homes from incoming Soviet missiles. Although the rainbow image may have been ill conceived—after all, Judy Garland immortalized it as a symbol of wishful thinking—it also evoked God's promise for protection against an apocalyptic flood from the sky. By association, SDI was an omnipotent guardian of the high frontier.

Critics of SDI countered with their own archetypal commercial. As a child stared at the night sky and sang "Twinkle, Twinkle Little Star," a point of light started to move and then zoomed down from the heavens. The most obvious and immediate objections to SDI were based on political considerations. SDI opponents charged that it would increase the risk of nuclear war. Unless the system was an absolute and perfect defense, they said, it automatically would be part of an offensive strategy and a threat to deterrence.

The rainbow started to pale when Reagan's original promise to protect population centers was downgraded to a plan for protecting missile silos. The National Security Council Decision Directive 172, issued on May 30, 1985, quietly confirmed the switch in

---

[1] See Jonathan Schell, *The Fate of the Earth* (New York: Knopf, 1982). On public anxiety in the early 1980s see Weart, *Nuclear Fear*, pp. 375–88. The Reagan administration's commitment to civil defense and nuclear war fighting capacity is described by Robert Scheer in *With Enough Shovels: Reagan, Bush, and Nuclear War* (New York: Random House, 1982).

thinking by redefining the SDI program as part of the strategy of mutual assured destruction.[2]

Weapons that can destroy other weapons, even imperfectly, might easily invite a preemptive attack. That means the weapons intended as a charm against intercontinental ballistic missiles might make us more vulnerable to war. The United States has never declared a "no first strike" policy. Even if we did, no rational adversary would trust such a pledge. A rational adversary would build up its own cache of ballistic missiles to improve the odds that its bombs would penetrate its opponent's space shield. For these reasons both superpowers agreed, in the 1972 Antiballistic Missile (ABM) Treaty, to forswear antimissile missiles. But research on how to hit a bullet with a bullet was permitted under the treaty.

The SDI proposal exhausted the domestic opposition to existing weapons, in the words of Spencer Weart, "by diverting it into a sterile debate over weapons that were not yet even on the drawing boards."[3] In truth, some of those weapons *were* on the drawing boards at the weapons laboratories, if only as crude sketches and vague ideas. The antinuclear movement lost momentum in the furor over SDI, but perhaps the greatest casualty in that political battle was the credibility of the scientific community.

SDI technology would be unbelievably complicated. Scientists skeptical of technical fixes for the arms race raised objections to plans to deploy an arsenal of smart rocks, brilliant pebbles, and lances of X-ray and laser light energy. But politics and skepticism aside, the complexity of SDI technology made it attractive. The technical problems had a sensual allure.

"I don't get my kicks from administering either projects or people," a physicist named Bobby said.

"And I'm frustrated by those idiots in Washington." He had just

---

[2]Richard Halloran, "U.S. Studies New Plan on Nuclear War," *New York Times*, May 25, 1985, p. 6; "NSC Quietly Rewrites Reagan's Star Wars Policy," *Newsday*, reprinted in *Albuquerque Journal*, June 15, 1985, p. B-9.

[3]Weart, *Nuclear Fear*, p. 385. On the political opposition to SDI, see George W. Ball, "The War for Star Wars," *New York Review of Books*, Apr. 11, 1985, pp. 38–44.

returned to Los Alamos from a trip to the capital, where he had tried to get research money from the SDI Organization's (SDIO) planners.

"The jargon term nowadays—I think we're called 'advanced concepts,' but that basically means 'crazy ideas,'" Bobby said, describing the mission of the twenty scientists he helped supervise in X-division. Some of those "crazy ideas" had meant Star Wars even before the president gave his famous speech on March 23, 1983. The group's projects included White Horse, the neutral particle beam intended to shoot down Soviet missiles. For five years one of Bobby's coworkers watched the group's funding shift and grow. At first about 20 percent of its projects were defined as weapons work. By the mid-1980s, 80 percent of its research was supported by weapons funds. SDI was the cause.

"They're muddled back there," Bobby declared, shaking his head at the memory of a trip he feared had been a waste of time. "Basically, Ronald Reagan says, 'Do Star Wars,' and someone says, 'Yessir!' They haven't the foggiest idea what to do."

Sandia National Laboratories, where nuclear devices get turned into nuclear weapons, has a secure niche in the division of labor among the weapons laboratories. Design assignments at Los Alamos and Lawrence Livermore are somewhat less definite. Those two laboratories are rivals. Each courts military and DOE planners with research and development proposals. (About two-thirds of America's stockpile of nuclear weapons was designed at Los Alamos.) After Reagan's Star Wars speech, fierce competition developed for the favor of the SDIO.

In the first year of Star Wars funding, Los Alamos took home $86 million for basic research on SDI. Sandia got $46 million. Gerald Yonas, a Sandia director, was given a leave of absence from the lab to become SDIO's first chief scientist. But Lawrence Livermore won the biggest prizes and was being publicly identified as the "Star Warriors" home base. When scandal hit SDI, Livermore bore the brunt.

If the rainbow was a false promise, then researchers who took SDI money were con men. That unpleasant charge originated with a computer specialist who designed battlefield management programs for the military.

David Lorge Parnas is a short, middle-aged redhead with a paunch and a barely visible pinkish mustache. His eyes are distorted by the kind of thick black-rimmed eyeglasses you might expect to find on the nose of a computer whiz. In mid-1985 Parnas resigned from the SDIO panel responsible for identifying the computer science requirements for a comprehensive ABM system.

Computers would actually aim and fire the system of SDI weapons. Like the string through a strand of pearls, computers would tie together the proposed array of orbiting high-tech baubles and trinkets. Parnas concluded that it was impossible to be sure the gigantic computer codes (with perhaps one hundred million separate lines of instructions) would reliably control the SDI hardware in case of a real ballistic missile attack. In numerous articles and lectures around the country, and to a very quiet crowd at a public colloquium at Sandia National Laboratories in 1986, this unlikely martyr explained why he had resigned his $1,000-a-day government consulting job.

In a nuclear war, Parnas argued, the computers would not have time to juggle their information and coordinate their instructions to the robotic guardians in the sky. Furthermore, there was no realistic way to test the computer brain of SDI. That is true of many technical inventions, but Parnas thought it the critical flaw in SDI. "If the first real use has to be successful," Parnas told the Sandia audience, "you can't do it." And a "probably perfect" strategic defense system could not be trusted. No policymaker could afford to rely on an untested SDI system to protect cities and missile silos. As soon as an attack was detected, a rational decision maker would order his own weapons out of their silos. Nuclear war would not be hindered.

Parnas warmed up the Sandia audience by joking about peaceniks and how he wasn't one. He had no moral objection to weapons work. But Parnas did object to projects that were useless, harmful, or fraudulent. At Sandia, his audience objected to the suggestion that they were wasting their time on SDI. Many asked questions like "Why not see the 'scientific impossibilities' as 'technical challenges'?" After each query Parnas would nod, thumb through his stack of transparencies, and slap one or another onto the overhead projector before giving detailed answers to which no Sandian offered a response. Parnas had yet to learn of an effective antidote to his technical pessimism.

Parnas first smelled danger and corruption when other SDI consultants argued that SDI should be pursued simply because it was a good source of research funding. He was not alone. By 1986, in an unprecedented display of unity, more than 6,500 university scientists and engineers—normally an individualistic and money-hungry bunch—had pledged to boycott Star Wars research funds. Signatories to the boycott pledge included over half of the faculty members in the country's top twenty physics departments. Among the petition's fourteen sponsors were five Nobel laureates and Carson Mark, a former director of the Los Alamos T-division.[4]

Accusations of scientific fraud became more specific as SDI researchers faced increasing pressure to demonstrate the potential of their programs. Positive results from a secret test at Lawrence Livermore in March of 1985 of the X-ray laser (the centerpiece of the proposed arsenal of antiweapon weapons) were "leaked" by

---

[4]Quotes from Parnas's lecture, "Why We Would Never Trust the SDI Software," Sandia Colloquium, Sandia National Laboratories, Aug. 1, 1986. Parnas's technical objections are in "Software Aspects of Strategic Defense Systems," *American Scientist*, vol. 73 (Sept.-Oct. 1985): 432–40. He has stated his ethical objections in numerous places, including "SDI: A Violation of Professional Responsibility," *ABACUS*, vol. 4, no. 2 (Winter 1987): 46–52. A response to Parnas's charges of ethical impropriety in SDI research is in Parnas and Danny Cohen, "SDI: Two Views of Professional Responsibility," University of California Institute on Global Conflict and Cooperation Policy Paper no. 5 (1987). See also Herbert Lin, "The Development of Software for Ballistic-Missile Defense," *Scientific American*, vol. 253, no. 6 (Dec. 1985): 46–53. Early criticisms of SDI's hardware are summarized in Hans A. Bethe et al., "Space-based Ballistic-Missile Defense," *Scientific American*, vol. 251, no. 4 (Oct. 1984): 39–49. Information on the scientific and engineering communities' boycott of SDI is from "Re: Boycott of Star Wars Research Funds by University Scientists and Engineers," public letter from Lisbeth Gronlund (Physics, Cornell) and David Wright (Physics, University of Pennsylvania), June 20, 1986.

At the University of New Mexico, which has close ties to Sandia and Los Alamos, the Faculty Senate passed a resolution prepared by the computer science, physics and astronomy, and mathematics faculty urging the university community "to refrain from participation in this ill-conceived venture."

Edward Teller despite Teller's knowledge that the X-rays had made the measuring instruments unreliable. Within two years, Lawrence Livermore Labs was in an uproar. Roy Woodruff, the former director of the laboratory's weapons program, claimed the Reagan administration had been systematically misled by Teller and others at the lab about the Super Excalibur X-ray laser program in a deliberate effort to make it seem more advanced and thus more deserving of funding. Then-Secretary of Energy John Herrington created his own minisensation by asking weapons lab scientists to keep their dirty linen off the public line. He feared the disputes would be "exploited" by SDI critics.[5]

As the Reagan era drew to a close the SDI program also faded away. Gorbachev seemed to offer hope for the redemption of the "evil empire" that the president had joked about bombing a few years earlier. The Los Alamos mathematician who had run in horror from Agnew's speech on the fortieth anniversary of Hiroshima had seen Star Wars as an invitation to nuclear escalation. By 1989 Sigmund had lost his fear. "SDI is dead," he declared flatly. The diagnosis may have been premature. Money for SDI research continues to flow to weapons labs and private contractors in a stream only somewhat constricted by Congress.

Much of the research funded by the SDIO had been cultivated in the weapons laboratories for years. The influx of money redefined old "basic research" as "national defense research" and allowed some lucky scientists to get their projects off the ground. For example, Los Alamos's Beam Experiment Aboard a Rocket promised to demonstrate a one-and-a-quarter-ton neutral particle accelerator from 125 miles in the atmosphere in April of 1989. By June the test had been delayed twice, the BEAR having eaten

---

[5]William J. Broad, "Science Showmanship: A Deep 'Star Wars' Rift," *New York Times*, Dec. 16, 1985, p. A-1; Broad, "Dispute on X-Ray Laser Erupts at Weapons Lab," *New York Times*, Oct. 22, 1987, p. A-11; Scheer, "The Man Who Blew the Whistle on 'Star Wars': Roy Woodruff's Ordeal Began When He Was Told to Turn the Vision into Reality," *Los Angeles Times Magazine*, July 17, 1988, p. 7; U.S. General Accounting Office, USGAO/NSIAD-88-181BR, *Accuracy of Statements Concerning DOE's X-Ray Laser Research Program*, June 1988; and "DOE Warns Scientists' Squabbles Damage SDI," the *Los Angeles Times*, reprinted in *Albuquerque Journal*, July 23, 1988, p. B-4.

about $50 million in SDI money. Other scientists at the weapons laboratories shook their heads in amazement as they saw their lonely little research projects suddenly shine in the constellation of SDI projects.[6]

Researchers at Los Alamos and Sandia had mixed feelings about SDI from the beginning. When the president first announced his Strategic Defense Initiative, some saw it as Reagan did, as a lofty guardian who could shelter us until it became obvious to all that nuclear weapons were obsolete. Most were noncommittal. One of the physicists in Bobby's crazy-ideas group counted the votes on his fingers. Six scientists supported the concept of SDI. Four expressed reservations. Ten were mum.

Bobby was one of the six SDI supporters in his Los Alamos group. At the end of the rainbow he saw a pot of gold. His support was contingent on the belief that SDI would probably never be used as a defensive shield in a nuclear war. Bobby said it was easy to decide if his weapons work was morally justified: "You're working on a given project and you ask yourself, 'If this succeeds, will I be contributing to another arms race?'"

Bobby had not even bothered to consider what would happen if Star Wars *did* work. "At that point—it's not clear where it is in my mind," he said, "but there is a line where I would probably have to look for work elsewhere. It seems fuzzy when posed as a hypothetical problem." He was annoyed by the suggestion that he apply to SDI the same hypothetical thinking that so engaged him with technical problems. "If you wish to engage in self-flagellation," he snapped, "you will concoct a scenario in which what you are doing is pivotal in an arms race, humanity goes down the tubes, and all that."

•   •   •

A clean-cut young physicist had gone to Sandia National Laboratories with the explicit understanding that he would not have to work on weapons. In graduate school he had been attracted to

---

[6] For a basic description of the BEAR project see Spohn, "Light Beam Meant to Zap Enemy Warheads," *Albuquerque Tribune*, Feb. 21, 1988, p. A-1.

fusion research for much the same reason as his fellow students. "People who work on fusion feel like they're doing a service to mankind," he said, leaning forward to sketch a TOKOMAK on a piece of scrap paper and digress into an introductory lecture on fusion. It promised clean, cheap energy. He remembered something else, too. "I had a very strong attitude about weapons-related work of *any* sort, including the kind I'm doing now." He thought it a waste of time and money, like war itself.

What had happened to his ideals in the four years since he had completed his Ph.D. and gone to work at Sandia? His interpretations had changed as the competitive egoism of the ace graduate student slowly dissolved in the bureaucratic environs of the weapons laboratory. "Even if somewhat highly trained and highly specialized," he had decided, "we're still just workers. We have very little control over how our work is used. I have very little impact on how the Star Wars things will be used in national defense. I'm not creating situations, I'm reacting to them."

The predicament of the moment was a shift in his group's research goals and programs. His fusion research was being tied to SDI in the lab's funding requests to Washington, and he was being asked to work directly on defensive weapons. The previously invisible line between offensive and defensive weapons suddenly started to emerge in his mind. It happened when his supervisor decided that everyone should have a "willing attitude" toward weapons work. Those who expressed reservations about SDI were labeled "potential defectors" from the division and suffered accordingly.

"I always wonder if I'm justifying my position—you know, rationalizing it," he said, his hands spread flat on the table between us. "I think most of us tend to work on the things that we're good at, and we justify that they're worthwhile one way or another." He and his wife were absorbed with their first child. Despite his busy schedule and new baby, he continued his volunteer work as a Cub Scout leader. "The Cub Scout thing is maybe a defensive mechanism for me," he admitted at the end of our discussion. It kept him in touch with individual human beings, he explained, and made him feel good about himself.

But he had lost track of the debates on SDI and arms control.

It made him uncomfortable, but he had decided to assume that our national leaders knew their business and acted in the public interest. In that light, the young fusion physicist was trying to think about those "requests" to work on SDI as "opportunities."

• • •

Duke may be the only Los Alamos staff member who wears a navy blue polyester blazer to work. The dapper sixty-year-old World War II veteran was a technician at the lab for ten years before he realized that he had more responsibility than status. The light dawned when he tried to attend a classified colloquium about the Strategic Air Command.

"I went to my supervisor and said, 'I want to go to this,'" he recalled over coffee in the lab cafeteria. "And he said, 'You can't, you're not a staff member.'" That policy still holds at Los Alamos. Duke was furious. He had more clearance than any staff member. He could go anywhere. His job as a plant engineering specialist required that he stick his nose into every nook and cranny of the laboratory. Duke's battle with the administration resulted in his eventual promotion, making him one of the few to penetrate the ranks of the Los Alamos staff with only a high-school degree.

Duke explained that he had not really understood about the atomic bomb until twenty-five years after the bombings of Hiroshima and Nagasaki, when he went to work at Los Alamos and started reading up on it. As he was talking his beeper went off—a problem somewhere in the lab. Excusing himself, he hurried out of the lunchroom and down the hall to the phone. A few minutes later he reappeared and resumed his train of thought. "It's my feeling that this laboratory should be all weapons and defense."

The major problem with the lab, Duke thought, was that it had a poor public image. Taxpayers saw it as a waste of money. "But you know," he countered, "there's some fantastic things that have come out of this—cancer research, solar research. Even though some of it has failed, it goes to the outside world." He listed more of the "spin-offs" that were part of the laboratory's technology transfer program.

If the lab should be completely devoted to national defense, I asked, why did he cite the "good things?" Was it hard to be proud of the weapons work itself?

Duke looked uncomfortable. "People get killed." He hesitated a moment, thinking how to explain himself. "Let's go back," he said. "The only one working in my family during the Depression was my grandfather. My father was out of work. He'd shovel a ton of coal for twenty-five cents to make money." Duke told me the whole story. As a kid he walked down to the railroad tracks every day with a little basket. The bits of coal he scavenged from beside the tracks fed his grandmother's cookstove.

"Times were tough, and people who didn't live through it don't understand," he said. "You know, bombs and weapons . . ." He stopped himself again. "As I say, this country has never had trouble when it was strong, only when it was weak. This country hasn't really been in a tough time since we developed the weapon."

Nuclear weapons and the Great Depression converged in his mind like parallel lines in non-Euclidean space. The bombs were a cellar full of hard black coal. Duke had no worries about nuclear war. "The leaders are just too smart to destroy the world," he assured me. "It's the terrorists we have to worry about."

•   •   •

A technician in the Los Alamos plutonium processing facility blows his nose and tosses the tissue into a trash can. At the end of his shift he strips off his gloves and steps out of his paper booties. The tissue, gloves, and booties contribute to the increasing stockpile of government-owned radioactive garbage.

Most of the 250,000 cubic meters of radioactive waste accumulated since the beginning of the Manhattan Project is buried in shallow trenches. It is difficult to retrieve because the soil surrounding it has also become radioactive. It will stay put for the foreseeable future, unlike the 57,359 cubic meters of transuranic waste stored in warehouses in Los Alamos and nine other DOE facilities around the country since 1970.

Transuranic or "TRU" waste is composed of anything higher

than uranium on the periodic table of elements. That includes plutonium 238 and 239 and americium 241, which pose long-term health and safety problems because they have long half-lives. For example, half of the plutonium on earth today will still be around twenty-four thousand years from now. Like most other TRU wastes, plutonium emits great quantities of alpha particles. These subatomic particles cannot penetrate a sheet of paper but are dangerous if inhaled or ingested.

The DOE's warehoused TRU waste fills 246,000 fifty-five-gallon drums. (There are 7,292 cubic meters in Los Alamos.) Over the next twenty-five years, the DOE will amass an additional 604,000 to 854,000 barrels. At minimum, that's more than enough to provide a barrelful of radioactive gloves, metals, rags, and chemicals (some toxic) to every man, woman, and child now living in Las Vegas, Nevada; Madison, Wisconsin; Little Rock, Arkansas; Fort Wayne, Indiana; and Syracuse, New York. Assuming they don't want it, what should be done with it?[7]

Bury it forever in New Mexico, says the Department of Energy. The Waste Isolation Pilot Plant (WIPP) is a 2,150-foot deep salt mine in the southeastern part of the state, not far from Carlsbad Caverns and the Texas border. Westinghouse is the prime contractor on WIPP, the nation's first *permanent* disposal site for nuclear waste. Sandia supervises the research.

---

[7] Figures are from Department of Energy, DOE/RW-0006, rev. 4, *Spent Fuel and Radioactive Waste Inventories Projections and Characteristics*, Sept. 1988. Nuclear waste has been classified into six categories that do not necessarily correlate with the level of biological hazard. Some "low-level wastes" (which include cesium and strontium) are included with the TRU wastes. These will decay nearly completely within about five hundred years but pose health and safety risks because they are "hotter" gamma ray emitters and require special remote handling procedures to protect workers. High-level waste, mostly generated by commercial and government nuclear power plants, is extremely "hot" and must be shielded with lead. For a summary of the types of waste and a basic introduction to the nuclear waste disposal controversy, see Anne L. Hiskes and Richard P. Hiskes, *Science, Technology, and Policy Decisions* (Boulder, CO: Westview Press, 1986), pp. 89–107. Don Hancock from the Southwest Research and Information Center guided me to sources of information on the technical and political issues with the disposal of DOE wastes. I also relied on an unpublished paper by D. Bruce Martin, "The Nuclear State Gets Dumped On: Radioactive Waste and WIPP," May 9, 1988.

The southeastern corner of New Mexico has the kind of terrain that seems fated to be called a "wasteland." At dusk, Mexican free-tail bats in search of flying insects flutter from the caves around Carlsbad in huge throbbing waves. Spiders, scorpions, rattle-snakes, and other nightmare creatures lurk under rocks and creep and slither across the hot, dry sand. Gray mice and glassy-eyed jackrabbits are preyed upon by fourteen species of hawks, owls, and other raptors. When these beasts die a natural death they are eaten by bald-headed turkey vultures.

The National Academy of Sciences recommended salt mines for permanent disposal of nuclear waste in 1956. Salt seemed dry. It is also very plastic. In theory, as the mine slowly collapsed, the rock would smoothly encapsulate the 6.3 million cubic feet of nuclear waste buried at WIPP like pearl around a grain of sand.

The Energy Department's "fortress in salt" was authorized by Congress in 1979. But before WIPP construction began four years later, the U.S. Environmental Protection Agency (EPA) and the U.S. Geological Survey foresaw problems with burying nuclear waste in salt beds. "This was advertised as a great place to put nuclear waste, because it's dry," explained Pat, a young geologist who had consulted on the WIPP project. "The truth is that it's not dry. There's water there. And if you get enough water and enough salt, you can corrode anything, including radiation waste canisters." Because of water, future sites for government and commercial radioactive waste will be in volcanic tuff or some other type of rock, but not in salt.

The desert near Carlsbad is bone-dry, but underground water seeps and weeps and shines on the walls of the mine at the WIPP site. Water is a problem because, like salt, it flows. Nuclear waste–laden water might flow through cracks in the salt and contaminate the Rustler Aquifer above the underground mine. (The aquifer feeds the Pecos River.) Or a driller seeking oil or gas sometime within the next ten thousand years, after the nuclear waste site beneath the desert floor has been forgotten, might drill into the radioactive tomb. If the sand at the heart of the pearl is dry, the driller would probably not contaminate the surface. If the nuclear waste repository is saturated with water, the pressurized radioactive brine would shoot to the surface.[8]

Ten thousand years ago human beings began their first experiments with agriculture and the domestication of animals. Ten thousand years from now, who knows what people will be doing? The EPA requires that these barely imaginable descendants be protected from hazardous releases from radioactive waste sites.

In what the EPA calls "human intrusion scenarios," the hypothetical miner must be assumed, and the safety of the people on the surface ensured, before a permanent nuclear waste repository can be certified safe. The Department of Energy hopes to experiment with nuclear waste in the site for five years before proving it can meet the environmental protection standards. An official of the U.S. General Accounting Office, in recent congressional testimony, said that whether or not WIPP could ever meet the EPA standards was "an open question."[9]

"If we were allowed to prove it the way I think the thing ought

[8] National Academy of Sciences, *Report of Disposal of Radioactive Waste on Land*, (Washington: National Academy of Sciences, 1956); U.S. Environmental Protection Agency, EPA/520/4-78-004, *State Knowledge Regarding Potential Transport of High Level Radioactive Waste from Deep Continental Repositories*, April 1978; U.S. Geological Survey, Circular 779, *Geological Disposal of High-Level Radioactive Wastes*, 1978. The brine problem at WIPP has two sources: slow-moving water in the salt formation itself, and a large natural pocket of pressurized brine that underlies about two-thirds of the waste repository. Should a miner drill *through* the WIPP repository into the pressurized brine beneath it, the brine might pick up radioactive materials on its way to the surface. The water at WIPP became a political issue when the Scientists Review Panel on WIPP issued its *Evaluation of the Waste Isolation Pilot Plant (WIPP) as a Water-Saturated Nuclear Waste Repository* (Albuquerque, NM, January 1988), which contains a bibliography of references on water in WIPP. I was one of two political scientists on the panel, which was organized three years after I interviewed Pat and others from Sandia involved in WIPP research.

[9] U.S. General Accounting Office, GAO/T-RCED-88-63, *Status of the Department of Energy's Waste Isolation Pilot Plant*, Sept. 13, 1988, p. 13. The EPA standards for permanent disposal of radioactive materials are the only relevant federal regulations governing the site itself, since WIPP was specifically exempted from the provisions of the 1982 federal Nuclear Waste Policy Act. Those EPA standards were vacated by the First Circuit Court of Appeals in 1987 on the grounds that they were not stringent enough, but the DOE and the state of New Mexico have agreed that the old standards apply until new EPA regulations are promulgated.

to go," Pat told me in 1984, "if we had another couple of years and a couple of graduate student researchers and all that jazz, I think we could show beyond a shadow of a doubt that it's a safe place for a repository of nuclear waste." Pat saw problems with the project, though.

"I don't think that most of the people I work with form their opinions on the basis of scientific fact," Pat explained. "It's become too political." There was a long pause, and then a reluctant conclusion. "I'm really making it sound awful. But I guess in fact it really is. There are a lot of people, especially on the engineering side, that just want to see it built."

The WIPP site was political from the beginning. Carlsbad, New Mexico, had two major sources of income: tourism associated with the hugely eerie Carlsbad Caverns, and potash mining. Exhausted mines and international competition closed the Carlsbad potash mines in 1968. With plenty of salt and unemployed miners, Carlsbad seemed an ideal nuclear waste disposal site. WIPP promised 2,150 initial construction jobs and another 950 positions for plant workers over the project's twenty-five-year operational lifespan, scheduled to begin in 1989. The five hundred to six hundred construction jobs that have actually materialized have been a lifeline for the desperate community. And the state of New Mexico has always welcomed atomic scientists and their projects.

Until WIPP. Opposition began in the late 1970s when a few activists realized that truckloads of nuclear waste would be rumbling down New Mexico's cracked highways to the nation's first permanent nuclear waste site. The "Citizens Against Radioactive Dumping" (CARD) were treated as crackpots and then ignored by the press, even after they tried for a more upbeat image by changing their name to Citizens for Alternatives to Radioactive Dumping.

CARD tended toward what one Indian friend of mine called "the Mother Earth, Father Sky approach." It carried little weight against the scientific details presented by confident and knowledgeable representatives of DOE and Sandia Laboratories. Those details were not always up to snuff, another Sandian confided. She had been invited to work on a program modeling the biological effects of an accidental release of radiation from the WIPP site.

The model was so shockingly unsophisticated that she thought it deceptive and refused the job.

The Southwest Research and Information Center, which analyzes environmental problems in the region, started monitoring the WIPP project and eventually became the semiofficial reliable representative of the "other side" in press reports and briefings of public officials. That group and CARD were the lone voices of public opposition for nearly a decade. The state government negotiated regulatory rights with the DOE after a 1981 lawsuit. Those negotiations were complicated, and the WIPP news confused or bored most citizens until 1987, when the governor suggested that WIPP should serve as a repository for commercial high-level waste. In exchange, Garrey Carruthers (a former assistant to Interior Secretary James Watt) hoped that the $62 billion federal Superconducting Supercollider would be built in the Land of Enchantment.[10]

Suddenly state residents started to worry about the fortress in salt. High-level waste sounded dangerous, and New Mexicans felt they owed nothing to the commercial nuclear power industry. A dozen independent scientists, engineers, mathematicians, and public policy specialists (including myself) formed a group to study the WIPP project. Using reports provided by the state's watchdog, the Environmental Evaluation Group, the "Scientists Review Panel on WIPP" eventually issued several analyses arguing that despite reassuring press releases from officials with Sandia and the DOE, the Carlsbad site was not ideal.

By themselves, neither the scientific debates about salt and water nor the fate of people ten thousand years in the future had much impact on public attitudes. However, the arcane controversies did make people throughout the state look out their windows and wonder what would happen when there was an accident with one of the flatbed trucks scheduled to bring between four hundred and nine hundred shipments of radioactive waste into the state every year for twenty-five years.

Sandia engineers had been working for years to perfect giant

---

[10]Carruthers was following the lead of an unsigned "White Paper" circulated among state government officials in mid-1987. A former Sandia administrator later

reusable metal casks to transport the fifty-five-gallon waste drums. Periodically the newspapers would announce a Sandia test of the strength of the TRansUranic PACkaging Transporters (TRUPACT). Front-page headlines about dropped, burned, and tortured but still-intact TRUPACT prototypes were often followed weeks later by smaller stories reporting that more modifications were needed before TRUPACT would meet Nuclear Regulatory Commission certification requirements.

The only reasonable route between Los Alamos and Carlsbad goes through Santa Fe. Specifically, it goes down Saint Francis Drive, a frantic six-lane highway that runs by a shopping center and blocks of older adobe homes. Part of the WIPP payoff to New Mexico is money for new roads, but the proposed bypass between Los Alamos and U.S. Highway 285 southwest of Santa Fe goes through San Ildefonso Indian land and old ethnic neighborhoods on the outskirts of the capital city. Until those people are convinced to accept the new bypass, Saint Francis Drive is the designated WIPP route.

Some anonymous public relations genius put homemade signs in the front yards on Saint Francis Drive announcing "WIPP RTE" in uneven red lettering beside three triangles in the classic configuration for radioactivity. Residents of "the city different" blinked twice and organized. When DOE representatives held hearings on revisions to its 1980 Federal Environmental Impact Statement in mid-1989, about 200 people in Albuquerque, and more than 550 in Santa Fe, trooped to the microphones. They denounced the highway plan; TRUPACT; the emergency training for local police, firefighters, and medical personnel; Sandia's reassurances about brine in the site; the proposal to put waste into WIPP for experimental purposes before meeting EPA standards— in short, everything about WIPP and the DOE that one could possibly imagine.

Some wore baseball caps advertising the "Worst Imaginable Potential Poisoning." Signs in the audience proclaimed "DOE = Death on Earth." It was the largest display of public hostil-

admitted to coauthoring the proposal. The superconducting supercollider project went to the state of Texas.

ity toward the nuclear establishment ever seen in the home of the bomb.[11]

•   •   •

When Dorie Bunting was fifteen, her parents sent her to live with a family in Germany for six months. They wanted her to learn the language. "That was in 1938," chuckled the pale-skinned, pale-haired Quaker woman. "That was probably the most critical experience in my life."

While Bunting learned German she watched people deny their most human instincts in a nation held captive to itself. During the Vietnam War she began to apply the simple principle that stayed with her long after the German vocabulary had faded from memory: One should act quickly when people start moving toward authoritarianism, violence, and other evils.

Putting her physical presence on the line seemed the strongest way she could send a message to the state. Bunting's first act of nonviolent civil disobedience was outside the WIPP site. She was among a small group of CARD protesters who blocked the road to the site in 1981. An old hip injury had left her with a visible limp. The police roughed up some of the demonstrators but allowed the tall, big-boned widow to walk to the police car. She spent three days in the Eddy County jail. "I was pretty scared beforehand," she remembered, "but it turned out not to be any great deal." Nonviolent civil disobedience became a habit.

Two years later Bunting and her friend Blanche Fitzpatrick opened the Albuquerque Center for Peace and Justice. Public demonstrations in Albuquerque usually turn out about one hundred protesters (objecting to a visit by Oliver North, for example, or U.S. aid to the contras in Nicaragua). The regulars worry if

---

[11] Within weeks Energy Secretary James Watkins announced that safety had become the DOE's number one priority and that therefore the opening of WIPP would be delayed until all the scientific questions had been answered satisfactorily. The original two-volume Federal Environmental Impact Statement (DOE, 1980) was supplemented by an additional two-volume Supplementary Environmental Impact Statement (DOE, 1989, draft) and is the most comprehensive source of information about the site.

Bunting is not there. Out of earshot, they call the woman who
rarely frowns or completes a sentence without a giggle the mother
of the peace movement.

"People become enslaved in their patterns," she said, talking
about the weapons engineers at Sandia Laboratories. "I understand that people have to have a livelihood and it's not all that easy
to find a job." We were in a lounge at the University of New Mexico. She had brought a stack of the Peace Center's newsletter to
distribute on campus.

Bunting tried to hold the other end of Chuck Hosking's banner
outside the gate to Kirtland Air Force Base at least once a week.
"So he won't get lonely," she explained. I had seen her there in
her usual attire: a plain cotton skirt and blouse and sensible shoes.
She would grin and roll her eyes whenever people screamed ob-
scenities out their car windows.

"At least they're thinking about it," she said. "It's a break in
their routine. They've been disturbed. The deadly routine . . ."
She shook her head and jabbed at her eyeglasses, which had
slipped down to the tip of her nose. "The automatism that sets in
with all of us, with the daily routine—it's *hard* to reassess what
your life is all about."

The weapons laboratories made it even harder, she thought.
Inside the fence and up on the Hill, the people who loved the
exciting technologies saw only a "monolithic affirmation" of their
work. Bunting thought she understood how it worked, but still it
made no sense to her.

"To contemplate using these weapons is beyond anything that's
rational," she said. "It's not sane. You wake up in the morning and
it's *beyond belief* that we've gotten into this situation. When you
think of four million years of human development—and we're willing to risk that?" When she laughed and shook her head her
glasses slipped again. She poked at them absently.

• • •

"I'm very selfish," Connie admitted. "If a bomb wiped out me,
my husband, and my child, I don't really care what happens to
anybody else." Connie had been designing thermonuclear weap-

ons at Los Alamos since she was twenty-five. The Personnel Office had invited her for an affirmative action interview in the late seventies. "I was lucky," she said, remembering the desperate scramble for jobs among the other physics graduate students. "There were people in my department who would have given their right arm just to get an interview here. But I firmly believe that, although I may have been initially contacted for affirmative action, I was hired because I'm competent for the job." When she finished her Ph.D. dissertation Connie became one of three female bomb designers in the X-division at Los Alamos.

The casually dressed young woman might have been any pleasant and competent professional. Her days were spent in an un-air-conditioned office on the second floor of the Los Alamos administration building. There 99 percent of her time was consumed by her "intense relationship" with a computer terminal. On the green screen she specified the ideal configuration of materials in a thermonuclear weapon—"What the thing would look like if the angels made them," she said, cribbing a phrase from one of the senior designers. Modifying an old weapon took a little over a year, Connie explained. Designing a new one took three years. When she wasn't working on the bomb, Connie fooled around with a little SDI research, but she didn't count that as weapons work. And she did not see herself as a relay runner in an arms race.

"Boy, if this was an arms race, it's not the way I would run one," she said. Resources were scarce, she explained, and there were too few people working on weapons design. In the previous five years her X-division group had had to replace thirty scientists and engineers. Only one had retired. The others had all burned out from the weapons design club's compulsion to meet the scheduled deadlines for tests at the Nevada Test Site. But Connie loved her work. "Being a peeping tom on Mother Nature is what it is," she said. "I get a big kick out of seeing how the physics is working inside the device."

Connie grew up in the South during the Cuban Missile Crisis. She remembered the duck-and-cover drills. The children in her elementary school were all fingerprinted so their bodies could be identified in case of emergency. (My mother had metal identification tags made for everyone in the family. We pinned them to our underwear every morning.) Nuclear weapons protests struck Connie as a fad. The antinukes, she explained, were in it for the fun of it. "It's a thrill to be afraid of nuclear war," said Connie. She thought those same people used their fear as an excuse for taking drugs and acting irresponsibly.

"All the antinuke people ever talk about . . ." Connie's voice got higher. She grimaced and rolled her eyes. "Is there *anything* that these people think is worth fighting for? There's something I'm willing to fight for. They say the worst thing that could happen would be a nuclear war. Well, I disagree with that."

Not that Connie saw the Soviets as the incarnation of pure evil. "I suspect that the majority of the Soviet people just want to have their house, their job, their car, uh . . ." Connie caught herself, apparently recalling that privately owned cars are few and far between in the Soviet Union. "Car" became "cat" in half a breath.

"And just to be left alone," she concluded. But the Soviet leaders struck her as fanatic, and she was afraid some of them might do something stupid in their obsessive determination to spread communism throughout the world. Weapons designers were like policemen, she thought. They help keep the peace. They may sometimes have to shoot people.

"Now, if there *was* a nuclear war, full-scale, I don't know how I'd feel. I certainly wouldn't be happy, but I wouldn't be happy no matter what I did." The bomb was mainly a design challenge for Connie. But it also shimmered like a mirage in her daydreams.

"Every once in a while, if my blood sugar's real low, I'll think, my God, my daughter's at home. If there was a nuclear war, how would I ever get to her? I've had nightmares about being separated from my family. But that's not related to nuclear weapons or anything." Connie had suffered from vivid nightmares about nuclear war and all kinds of horrible things long before she went to Los Alamos. She thought they had no special significance. "I'm weird that way," she shrugged. "I don't know what it is."

12

THE

DISENCHANTED

•
•
•

The information packet for newcomers from the Los Alamos Chamber of Commerce includes a mimeographed table that describes the ups, downs, and averages of the climate on the Hill between 1911 and 1981. The low temperatures in January averaged 18.3 degrees Fahrenheit; the highs, 39.76 degrees Fahrenheit. An average of 11.96 inches of snow falls each December. The precise figures have simple practical significance: mount heavy tires on your car and carry chains in the trunk. Since indiscriminate use of salt on the highways strains the fragile ecosystem, the county trucks spread sand for traction on the city streets. It stains the snow blood red.

Two thousand feet below, Albuquerque has milder winter temperatures and less snow. Only once or twice a year does it blanket the Sandia and Manzano mountains and stick all the way down to the crease of the Rio Grande Valley. Skiers rejoice. More than three or four inches on the city streets closes the schools and public offices, thus protecting residents from each other's Keystone

Kops—style of winter driving. The city usually returns to life within a few hours as snow at the lower elevations evaporates in the dry desert air.

Snow shines on the desert floor like salt crystals on creamy scrambled eggs. The limestone and granite mountains of New Mexico seem even bluer against patches of clear white. Every winter, the state National Guard air-drops emergency bales of hay to starving livestock on the Navajo reservation. Sometimes hikers in the high northern mountains get lost and die of exposure.

But the summers are more dangerous in the Land of Enchantment. The black widow spiders become active and leave festering wounds and high fevers with each tiny bite. Melting winter snow and rainwater from storms at the top of the mountains rushes right down the clay and rock in deep natural ditches. The larger ditches that run through cities are often paved, making them irresistible to kids on bikes and skateboards. Every spring some are surprised by a sudden swell of water in the arroyos, knocked unconscious, and washed downstream. As summer progresses, the forests turn to crispy kindling for fires started by lightning and careless campers. Rain clouds promise cool showers, but the drops evaporate before touching the ground. The midday sun makes people dehydrated, cranky, and sick.

This might help explain why employees of New Mexico's weapons laboratories weren't much concerned about the idea that a thermonuclear war might change the global climate. The image of smoke-filled skies over a cooled earth after a thermonuclear war was suggested in 1982 by Paul Crutzen and John Birks, in a study commissioned by the Swedish Royal Academy of Sciences. The following year, more elaborate models describing the effects of dust and smoke from burning forests and cities were published by five scientists—Turco, Toon, Ackerman, Pollack, and Sagan. The so-called TTAPS models simplified the complex natural movements of air around the globe. Even so, the predicted effect of nuclear war on the global climate was sobering, especially after biologists considered the consequences of low light and unseasonable cold on plant and animal life.

The earth would be abnormally cold and dim for perhaps as long as two-and-a-half years following the TTAPS scenario for a "base-line" war—five thousand megatons destroying 242,000 square kilometers of cities and suburbs. Survivors in both the northern and southern hemispheres would face food shortages that might persist for decades. Animals would be hungry, thirsty, poisoned, sick, and shivering. If destruction of the ozone layer allowed high levels of ultraviolet-B light to reach the earth, many would also be blind.

Most people at the weapons laboratories found SDI more interesting to contemplate than nuclear winter. Nuclear winter had Carl Sagan, but SDI had money. And Carl Sagan was easy to caricature. In the early 1980s he was America's greatest scientist-popularizer, trying to instill in the general public his own awe for the "billions and billions of stars" of which our own sun is only one. When Sagan stuck with glitzy introductions to physics and astronomy, people at the labs thought his efforts at public education commendable. When he publicized the nuclear winter studies in a cover story for *Parade* magazine, people like Connie thought him guilty of "criminal irresponsibility."

The weapons laboratories responded to the threat of nuclear winter by intensifying their research on "earth penetrator weapons." These bombs screw themselves into the earth before exploding. They would cause fewer fires and kick up less dust than ordinary thermonuclear weapons. Because they can also destroy hardened targets, like the enemy's underground military command centers, the military has been enamored of penetration bombs since the 1950s. The obsolete, pointy-nosed Mark 8 on display in the National Atomic Museum could penetrate 120 feet of clay, 90 feet of hard sand, 22 feet of reinforced concrete, or five inches of armor plate before it detonated.[1]

---

[1] Paul J. Crutzen and John W. Birks, "The Atmosphere After a Nuclear War: Twilight at Noon," in Jeannie Peterson, ed., *The Aftermath: The Human and Ecological Consequences of Nuclear War* (New York: Pantheon Books, 1983), pp. 73–96; originally published in *Ambio*, vol. II, nos. 2–3 (1982). The original TTAPS report, along with elaborations on the biological effects of climatic change, is reprinted in Paul R. Erlich et al., *The Cold and the Dark: The World After Nuclear War* (New York: W. W. Norton, 1984). See also Carl Sagan, "The Nuclear Winter,"

If "The Cold and the Dark" indecorously dramatized science and needlessly scared the public, many weapons scientists nonetheless thought a different scientific horror show long overdue. We have all become too casual about nuclear weapons, they said, the old films of atmospheric bomb tests and the names of Hiroshima and Nagasaki now little more than abstract symbols of the nuclear age. Why not demonstrate the power behind the symbols with a big aboveground nuclear test? Imagine yourself standing in a desert wasteland, they suggested, or on a ship in a remote area of the ocean. All the world leaders are gathered with you to observe the explosion ("From a safe distance, of course," one Los Alamos weapons engineer added quickly). The film becomes real as the world goes white. The supernatural cloud from a thermonuclear explosion rises and swells. The colors shift. You feel the heat. You hear the sound of distorted thunder.[2]

• • •

In 1975 Gordon McClure's son asked him, "How *many* of these nuclear weapons do we have?" McClure had been at Sandia since the 1950s and had supervised a division of weapons components researchers since 1969. He did not know the answer. Six years later, on the verge of retirement, he decided to find out.

The physicist found himself driven to study every conceivable strategic, military, political, and technical aspect of nuclear weapons policy. The demon for scientific details slowly started to see the big political picture. McClure began publishing guest editori-

---

Parade magazine, Oct. 30, 1983, pp. 4–7. The scientific controversies and policy implications are examined in "Nuclear Winter: Uncertainties Surround the Long-Term Effects of Nuclear War," U.S. General Accounting Office, GAO/NSIAD-86-62 (March 1986). The data on the Mark 8 is from Carson Mark et al., "Weapon Design: We've Done a Lot but We Can't Say Much," *Los Alamos Science*, vol. 4, no. 7 (Winter/Spring 1983): 161.

[2]The original proposal for public demonstrations of thermonuclear explosions has been attributed to Harold Agnew. Many weapons lab employees thought it a good idea—once every five years would do it, one said. None seemed concerned about the radiation and fallout.

als in the local newspapers. He gave over one hundred public talks in the first three years of his retirement. You could find him on the radio or sitting on a panel at a public forum, a thin man with a strong sense of irony revealed by a frequent, broad grin. When Sandia's laboratory colloquium committee unanimously proposed that McClure be invited to the lab to speak on his views, however, a vice-president vetoed the suggestion.

McClure was obsessed with the development of weapons components when he worked at Sandia. "You're so goddang busy from day to day that you put off questions about the implications of it," he remembered. The continual temporary management crises were frustrating distractions. McClure later compiled a list of the unexamined justifications he had accepted without question over the years. In retrospect, McClure figured that he and others at the weapons laboratories suffered from a double layer of psychic numbing, which is essentially a form of denial. Nuclear winter shook him up. "The worst thing," McClure said, "is that the idea came from *outside* the weapons laboratories."

It seems that no one paid to think about nuclear war had ever seriously considered the consequences on global climate. They might have come close. *The Effects of Nuclear Weapons* contains a brief consideration of the ability of different colors of smoke to reflect and absorb energy. The smoke might be released deliberately, either by a nuclear weapon or a previously detonated smoke generator. The unclassified Bible on nuclear weapons also mentions that "debris entering the stratosphere may interfere with the transmission of radiant energy from the sun and so serve to decrease the temperature of the earth." That effect was considered unlikely. (Richard Turco, a primary author of the TTAPS report, assured me that there was no additional elaboration on such matters in the classified literature.)[3]

---

[3] The concept of psychic numbing is developed by Robert J. Lifton in Lifton and Richard Falk, *Indefensible Weapons: The Political and Psychological Case Against Nuclearism* (New York: Basic Books, 1982). The quote is from Samuel Glasstone, ed., *The Effects of Nuclear Weapons*, rev. ed., prepared by the U.S. Department of Defense (U.S. Atomic Energy Commission, Apr. 1962, changes as of Feb. 1964), p. 85; see also pp. 321–23.

"What have we been *doing* all these years?" McClure wondered. He grinned and chuckled when he remembered his own frame of mind. "I thought about things like, 'Well, what kind of a weapon would you like?' "

Just before Christmas in 1985 McClure sent a "Dear Friend" letter to sixty people he knew from the laboratory. In it, he explained his change of heart and described the crucial irrationalities in the arms race. McClure promised to buy lunch for the first ten people who took up his offer to get together and talk about the issues. A year later, nine lunches remained unclaimed.

•   •   •

When the first generation of atomic scientists designed their bombs, they scribbled their equations on paper and chalkboards. The modern tools of the trade are quicker and more sophisticated. They are not perfect, however.[4]

"Most problems with the computer derive from the fact that it does exactly what it's *told* to do, not what you *want* it to do," Victor said. The forty-three-year-old Los Alamos engineer laughed with delight at the stupidity of his machine. For the next two hours he would talk nonstop and display no other sign of emotion.

One hundred thousand lines of computer instructions, give or take, define the guts of a thermonuclear weapon. Victor once wrote those mammoth programs. Tall, thin, and unearthly pale, his light gray eyes shifting erratically, his dry white hands vibrating on the tabletop, he might have been an android with jumbled circuits. He seemed to have abandoned sleep in favor of caffeine. He quivered and hummed like an engine on fast idle. His high-speed rattle turned the coffee-shop waitress's efficient bustle to slow motion.

Victor considered himself "a details person" and "a scholar." Designing the intricate logical mazes that nest like Chinese boxes

---

[4] For a brief history of computer use in weapons design, see Francis H. Harlow and N. Metropolis, "Computing and Computers: Weapons Simulation Leads to the Computer Era," *Los Alamos Science*, vol. 4, no. 7 (Winter/Spring 1983): 132–41.

inside the weapons' computer codes was perfect for him. But Victor had transferred out of X-division into an easier job checking other people's computer programs.

Weapons design work was too nerve-racking, he explained. "I have seen extreme pressure placed on designers to accept unsatisfactory construction or an unsatisfactory product in order to meet an artificially set schedule." Sometimes things were not built to spec. The "specs" for the "products" were engineering specifications for thermonuclear bombs. The "schedules" were deadlines for the underground test shots the designers try not to slip.

Details people were unappreciated, Victor thought. The "big-picture people" forgot the debt owed to those whose nails were bitten to the quick from worrying about how to fit the smaller fractions of reality into the larger abstractions. Some hands-on weapons designers loved the physical mechanisms of modern bombs, the parts and pieces, but not Victor. His bomb was an elegant set of logical instructions, his sanctum in Los Alamos a cramped office with a computer terminal. His keyboard served as compass and square.

Victor had no desire to see the cloud or feel the heat. He loved the logic and the numbers. He described his feelings about life in Los Alamos in a single sentence: "It's situated in the mountains, and I love the mountains." As I tried to visualize the ashen-skinned computer scientist hiking under a hot sun in a huge sky, he began to recite figures on the number of bookstores and movie theaters per capita in Los Alamos. (The Hill has more places to buy books and fewer places to see movies than is the norm for American cities.) The statistics grew more complex and arcane, and Victor grew more animated. I could imagine him as a white-water rafter hurling down a digital stream of consciousness.

After giving a statistical profile of Los Alamos, Victor concluded that he lived in a strange town. He set about explaining his social responsibility as a scientist with similarly complex indirection. First he reconstructed the natural and political history of the human species.

"Take the United States," Victor suggested, turning his head toward the window that framed a busy Albuquerque intersection where pedestrians skipped and weaved their way around unsympathetic cars.

Many of those people outside the window were pretty nice, Victor said. Some were not. "We call them 'criminals'," he explained. "To solve the problem of criminals, we arm people and call them 'police.'" Cops and robbers formed a system. Rules constrained American police and impair their functioning, but those flaws in the system could be fixed by careful social engineering, Victor said, which was why he supported the election of judges. He shifted topics while I filled in the blank: Elected judges are more likely to be hanging judges who can compensate for the system's imperfections.

The community of nations also required a policeman, Victor was saying. "Black Bart the Bad Guy" was the Soviet Union. With the United Nations unable to function as an international security guard, he thought it natural that the United States assume the function of maintaining world order.

But that, Victor said, led immediately to more fundamental questions, which turned out to be, "What are the legitimate functions of government?" and "What are the rights of man?" He answered with a vague reference to Thomas Jefferson and continued briskly through an immense array of large abstract theories and small detailed facts about politics and nature. The founding fathers took him to the Declaration of Independence, and then to the Soviet invasion of Afghanistan, the U.S. invasion of Grenada, the use of mustard and nerve gases by Iraq against Iran, the general characteristics of Third World revolutions, the U.S. mining of Nicaraguan harbors, the politics of President Eisenhower and Senator Kennedy, and the Magna Carta. He ricocheted from the Magna Carta to the army, recreational hunting, gun control, antiballistic missiles, and the bias in the American press.

Victor started describing the function of females in mammalian species and historical shifts in infant mortality rates. After two hours, his ideas seemed to be getting away from him. The details had distracted him from his original intent, which was to explain his social responsibility as a scientist. I interrupted to ask how he felt about his work.

Victor stopped dead. "When people ask me what I do for a living, and I tell them I help design atom bombs—like someone you meet on an airplane—normally they respond with horror: 'How could you do such a thing?' After the second time I formulated a response."

"We, the people of the United States, have been assured by our elected officials that they will use nuclear weapons only in defense of the country, and insofar as this is a legitimate function of government, mine is an honorable occupation." He recited it like an unloved poem dredged deep from memory.

But how did he feel about that? I asked. Victor slumped a bit. "I help design atom bombs. I'm only a small cog in a complex machine, but I try to do my job competently and earn my pay." He stared out the window. "I don't know exactly where along the line I got some of these hard attitudes toward life," he said. What did he mean? From inside the machine the civilized scholar suddenly turned and pointed an accusing finger at the great outdoors. "I mean," he replied, "Mother Nature is a mean bastard. She always collects. The only question is who pays and when. She always collects."

•   •   •

Civilization ends, wilderness begins, and real estate is notoriously expensive on the roads that cut across the western slope of the Sandia Mountains. Roberto's designer adobe home with passive solar features was at the top of one of the uppermost trails through the foothills. The small tan house had sand and scrub in place of the normal suburban lawn. The interior looked like a church, with skylights in the high ceilings that cast long moving shadows on the brick floors as evening became night.

The darkly handsome computer analyst was in his mid-thirties. Roberto had started out at Sandia in a general-purpose computing group. Then he transferred into a group that designed programmers—"the part of the fuze that coordinates what the other parts of the fuze do"—for the Trident II missile.

Sitting in a high-backed wooden chair at the heavy, Spanish-style dining table, Roberto said quietly, "Everyone participates in weapons work in a way. If you pay taxes you're supporting the weapons system. Am I more involved just because I use my expertise to build nuclear weapons?"

He answered himself with a smooth torrent of words. "A lot of the people you would talk to in the weapons building business would say they're the most socially responsible people around," he said. "They're helping protect us against communism. My own personal opinion is that when you're talking about bringing several civilizations to an end over petty politics, that's pretty self-righteous. But in New Mexico, that's the kind of work there is."

Roberto was guessing about his coworkers' reactions. "Rarely do we talk about this," he explained. The tradition of silence began in Los Alamos, in the hothouse of scientific discovery at the beginning of the nuclear age. John Manley had thought it a characteristic of his generation, raised to treat political opinions as private, "sort of like inquiring into your sex life or something like that."

Roberto thought his colleagues simply could not face the reality they helped create. Some joked about it. Others adamantly refused to discuss it and would not even acknowledge seeing the guy the jokers called "Bicycle Bob." Although he had never stopped to speak with Chuck Hosking, Roberto was always pleased to see him standing with his signs at the gate to Kirtland Air Force Base. "I would be very disappointed if he were gone and I knew he wasn't going to come back," he said. "He may be the closest thing we have to a conscience.

"I could wake up one morning and flip my badge over the fence," Roberto said in a level voice. "I could say, 'Hey, it's been great, but I don't want to be doing this.'" But Roberto was proud that his Hispanic family had lived in the area for four hundred years, and he would never consider leaving the Land of Enchantment. Sandia offered the most interesting assignments, the best working conditions, and the highest pay in the city. "One could argue that the ethical thing to do is not work in that situation, not contribute that expertise," he continued.

"Or one could argue that the practical thing to do is to maintain your skills and say, 'Hey, these are *defensive* weapons.' I'm not sure either is right." It had grown quite dark, but Roberto was staring down the length of the long carved table and did not seem to notice. He talked about the different ways one could make peace with oneself as a weapons worker. He thought they were all rationalizations.

"To be perfectly honest with you, I'm not encouraged for our prospects." His face and features in shadow, he said, "I'm glad that I don't have children."

•   •   •

Sandia's Tech Area One is enlivened by one small spot of green. A dozen or so trees grow from square planters on a square brick plaza near the laboratories' library. "They call this Sparks' Park," said Tom Grissom, smiling as he told the story. In 1970 Sandia had an unprecedented layoff of workers. Two years later, under the presidency of Morgan Sparks, the landscaping of the plaza reassured workers that happy days were there again. A second layoff a few months later instantly turned the park into a memorial to lost colleagues and broken promises.

Grissom loved Sparks' Park and made it the first stop on our six-hour walk through the labyrinth of buildings in Tech Area One. The tour was a sentimental journey for the forty-five-year-old experimental physicist. In one week he would be leaving the weapons lab.

Grissom was a calm southerner with classic, even features, bright blue eyes, and sandy brown hair just starting to thin and gray. He had left the University of Tennessee fifteen years earlier with his Ph.D. and two job offers. An avid archer, Grissom had managed to convince a tiny bow-and-arrow manufacturer that it might benefit from some sophisticated research and development. He had also been interviewed by Gordon McClure at Sandia. "I took the conservative yuppie way out," Grissom explained. "I really didn't think much about nuclear weapons at the time."

By 1985 Grissom was earning $75,000 a year managing seventy researchers and technicians in Sandia's Department of Neutron Devices and Technology, otherwise known as the "Tube Lab."

Neutron devices are the oblong initiators that trigger the chain reaction in a nuclear bomb by shooting a blast of neutrons into a mass of plutonium. Grissom attributed his success as a manager to his skills as an arbitrator, negotiator, and peacemaker. "I always liked people for what they were," he said. "But it doesn't make you as critical as you should be. You become accustomed to not saying what you really want to say."

It took him a while to realize that he felt alienated from his work. "I think I have known for so long that I had to leave Sandia that I can't identify with any accuracy or precision the point at which that decision occurred," he admitted. "At first the disagreement was just intellectual." Later, after a bitter divorce had made him reassess his life, Arthur Clark's biography of Einstein sent him in search of creativity and inspiration. Grissom read three hundred books in two years. Much of it was poetry. "Literature offers a whole world in a microcosm," he said, "a whole world of feelings." The abstract questions about the purpose of the laboratory gradually became questions about the purpose of his life.

Grissom began writing his own poetry, originally just to see if he could do it; later, it became a means of self-expression. One day he sent some off to a small press. To his surprise, the press offered to publish them.

We had wandered around and through the buildings in Tech Area One for five hours before Grissom led me into the office he was preparing to abandon. "No messages," his receptionist said cheerfully. The long wall of his big office was lined with a bank of locked tan metal cabinets.

Grissom sat behind his desk and stared at the expanse of cream-colored wall-to-wall carpeting. He seemed at a loss. He had given one month's notice, he explained. His calendar, normally dense with appointments and reminders, progressively and inevitably showed nothing but blank space. The past two days he had spent rewriting the introduction to his second collection of poetry. He pulled it out of his bottom drawer. "Viewed all together," it began, "these are poems of survival."[5]

---

[5]Thomas Grissom, *One Spring More* (Francistown, NH: Golden Quill Press, 1986), Introduction.

Grissom had spent a year searching for a research job. He turned down some good offers. "I decided there's really nothing I can do that's really clean, that has no military applications," he said. Then he answered an advertisement from Evergreen State College in Olympia, Washington, a small undergraduate school that took interdisciplinary studies seriously in the 1960s and never reneged. Grissom wrote that he was a physicist and a poet seeking personally rewarding work that had nothing to do with the weapons industry.

After Grissom decided to leave the laboratory but before he knew where he was going, he sat down "in a moment of great resolve" and composed a six-page essay written to explain himself to his children and a few friends. It began, "After fifteen years I am leaving Sandia. The reason is very simple—conscience and honesty demand it."

"In the main," he wrote, "we are engaged in building toys for professional soldiers who need them to justify their existence, and for an industrial society which requires them to provide jobs for its workers, at the bidding of politicians who lie and manipulate us and prey upon our fears for their own ends, in a vast and interrelated scheme reminiscent of a teeming anthill with its soldiers and workers and rulers, each carrying out its separate function in orderly obedience to genetically programmed behavior, unaware of why or even of the net effect of individual actions. For us the end result of our separate behavior could well be nuclear catastrophe."

"But we are not ants," Grissom continued. He went on to speculate about what might happen after a nuclear war. Suppose the survivors were to round up the weapons scientists and put them on trial for "heinous crimes against humanity"? "I prejudge our guilt to be the same as that of all the good Nazis who obediently followed orders," Grissom wrote. He entitled his essay "To a Few Friends," and when Evergreen State College wrote back asking why he suddenly wanted to quit his job, he sent them a copy. The college offered him a one-year teaching job at less than half of his Sandia salary. He accepted. Sandia offered him a one-year leave of absence. He declined. Just before he left Sandia, he made copies of his letter for everyone in his department. "They're going to think I'm a nut," he predicted.

In fact, Tom Grissom's decision to leave Sandia left his best friend in anguish. Besides being an engineer in the Department of Neutron Devices and Technology, Leon was an atheist, a strong believer that free will was mere fantasy, and a philosophical proponent of communism. Leon had light red hair that faded to white on the edges of his beard and mustache. "I'd be willing to give some of this up for the common good," he said, peering out from behind his thick horn-rimmed glasses and waving his hand around the room.

We were in his den, surrounded by the books that he constantly reshelved in an endless effort to find a rational system of classification. Should B. F. Skinner's *Beyond Freedom and Dignity* be placed with philosophy or psychology? Where does one end and the next begin? Leon loved to talk. Every new topic was an occasion for him to point to the books that had most affected his thinking. He lost his faith in God after reading James Frazier's *The Golden Bough* and found the virtues of communism compellingly described in *The Communist Manifesto*. *Small Is Beautiful* and *The Limits to Growth* made him believe that Americans had become unbelievably self-indulgent.

Grissom had explained the apparent unanimity of opinion at Sandia by sheer force of numbers. "There's eight thousand people who work out here," he had reminded me, "and if one stands up and says, 'This is wrong,' it's eight thousand to one. It's just not allowed."

Leon disagreed. He thought that people like himself represented a small but meaningful oppositional force within the weapons laboratories. "The people in the peace movement are better off having me in there than my replacement," he said. "The people who *would* quit are the ones you want to stay in there. I'm a dissenting voice and I'm not afraid to speak out." Some of his coworkers seemed unbearably foolish, like the engineer in the office next door who believed that the whole universe was created six thousand years ago. With no effort at all Leon could set off "little bombshells" in his conversations at work. All he had to do, he claimed, was to say what he thought.

Then Leon remembered that he was a strict determinist. "I

don't think I've influenced *anyone* in this life," he confessed as he propped his thin bare feet on the coffee table. What people did, they would do anyway. How they felt was irrelevant. How did he feel? He blurted out the answer. "If you're building bombs—I don't know—if you were building furnaces for Hitler, how would you feel?"

Leon was testing the neutron bomb when it was so controversial in the media. He thought it a terrible weapon that would make it too easy to back into a nuclear war. Unlike most of his colleagues, he was frightened by the scenarios for nuclear winter. "I think nuclear winter is a very real possibility," he said, "and it just kills me to think that man might continue, but that no one would ever read Emily Dickinson again, ever." He looked over toward the poetry section of his bookshelf. "It's just *appalling* to think of it."

Nonetheless, Leon loved his work. "It's hard for people outside to understand the seductiveness of this," he said. "You have very strong social encouragement and you have the toys to play with— you know, the instruments." In one month he would be eligible for retirement. The idea made him anxious. "What would I do?" he wondered. Leon had decided to stay on at the laboratory as long as he could.

"I know how I feel about nuclear war and yet I don't feel bad about what I'm doing," he said. "Ergo, I must be rationalizing. I don't know. I haven't deceived you. I haven't spoken one word tonight that I don't believe, the inconsistencies and all." Such inconsistencies, he thought, were nearly universal. They resulted from our stubborn attachment to the delusion that people have free will.

●　　●　　●

In the course of two years I interviewed eighty-five people who had worked in New Mexico's weapons laboratories and spoke informally to many more. I asked each if they had ever wanted to quit their job. Only eight said no. "I did, the day I came," said one physicist who had been at Los Alamos Lab for eight years. "It was the last place I thought I'd ever work. In the sixties, I was quite active, a radical." He was raised by religious pacifists. "I really didn't want to work here," he said. "It was a tremendous defeat."

A technician responded with bitter resignation. "Yeah," he said, "I feel guilty about being part of a damn group that builds destruction. I think it was the worst thing that ever happened, the bomb." But, he said, the money was too good to pass up.

Most would leave only if they could find something more interesting and rewarding. They had gone to the weapons laboratories expecting magic. Then they realized that their freedom was tenuous and constrained by both explicit and self-imposed rules and regulations. Their basic research was vulnerable to redefinition as an irrelevant waste of the taxpayers' money or potential military technology; their truly secret arts turned out to be stressful and unglamorous. Their fellow wizards, trained to be rational, creative, and objective engineers and scientists, were just as annoying and no more profound than ordinary mortals. People at Sandia found themselves enmeshed in a humorless bureaucracy. Those at Los Alamos felt isolated on the "Mesa of Doom." Disillusionment set in when it turned out that working in a nuclear weapons laboratory was just another job.[6]

Grissom is the only person I know of who admitted to leaving the weapons laboratories for moral reasons. He had felt compelled to the decision. "A thousand weapons could wipe out the world, if they were properly placed," he said. "We have thirty, thirty-five thousand—I could tell you the exact number, but that's really classified. I've never heard a manager say, 'Do we really need this weapons system? Is this really a good idea?' Maybe if I'd heard that I'd be willing to stay."

"I'm a pessimist," Grissom continued. "I believe nuclear weapons will be used. I can imagine situations in which they would be used. People who can't, I think, suffer from a terrible lack of imagination." He was rapidly losing patience with those people. He did not expect them all to quit their jobs, but he wondered how so

---

[6]The term *mesa of doom* is used by Salholz et al. in "An Identity Crisis at the 'Mesa of Doom,'" p. 30.

many could stand to trade prestige and a paycheck for hardened hearts and dulled senses.

"The honest truth is, I don't expect what I'm doing to have any real effect on anybody except that it will serve as a topic of idle conversation and gossip," Grissom said calmly. "The impact will be short-lived and totally insignificant. I'm only doing it because I'm just weary of the illogic of what we're doing; the danger; the absurdity. I just heard the person inward sort of saying, 'I'm not going to do this anymore.'"

EPILOGUE

The old one-lane Otowi Bridge crosses the Rio Grande at a place the Indians used to call Po-sah-con-gay, "the place where the river makes a noise." Where the river's shallow banks suddenly become steeper and the riverbed constricts, the muddy water roars and gurgles in tangled skeins through rocks and old tree roots. The bridge links the Parajito Plateau to the valley below. Across that narrow span Los Alamos scientists carried the pieces of the Trinity bomb on their way to the flat white desert on the Jornada del Muerto.

In 1947, when it became clear that Los Alamos would never again be isolated from the rest of the world, a new two-lane bridge was built at the Otowi crossing. The construction displaced a woman named Edith Warner. For twenty years she had run a small tearoom and supervised the station where mail and supplies bound for the Hill were stored until a truck could be sent down from the Los Alamos Boys' School. During World War II the elderly spinster was the atomic scientists' closest Anglo neighbor. Dozens of

her neighbors, scientists and Indians both, volunteered to build
her a new adobe house. She chose a spot a half mile down the
river, far from the noise of the cars and trucks that carried heavy
equipment, building supplies, and more scientists to the expanding weapons laboratory.

Warner had moved to New Mexico in 1921 with the excuse that
her fragile health required plenty of fresh air and few responsibilities. Beside the river the delicate daughter of a Pennsylvania Baptist minister learned, in the words of her biographer, "what it
means to live at the center of a sacred world." Many people have
come to New Mexico for that purpose. At the end of World War I,
profoundly disillusioned with Western civilization, D. H. Lawrence sought an antidote for his pessimism in the mountains of
northern New Mexico, where the Indians seemed to accept, understand, and accommodate the dark and primitive spirits that the
Europeans had tried to suppress. Lawrence died ten years before
the first atomic bomb was exploded. Georgia O'Keeffe came later,
seeking a place where life and land could be depicted as simple
forms in pure colors. Other artists, writers, and seekers of spiritual
truths have found the eternal quality of the sacred world somehow more palpable in the mountains and deserts of the Land of
Enchantment.[1]

The old Otowi Bridge is on land belonging to the San Ildefonso
Indians. The new two-lane bridge beside it was repaved and expanded to four lanes in 1989. Between the old and new bridges,
construction workers have erected a new fence. The four strands
of tightly strung barbed wire are already stretched out of shape by
people who have pulled them apart so they could climb through

---

[1] The story of Edith Warner and the scientists who had visited her at the Otowi
crossing was written by Peggy Pond Church, the daughter of Ashley Pond. See
Peggy Pond Church, *The House at Otowi Bridge: The Story of Edith Warner and
Los Alamos* (Albuquerque: University of New Mexico Press, 1960), p. 19. For
Lawrence's disillusionment with the postwar world and a description of the artistic
community nurtured by Mabel Dodge Luhan in the northern part of the state, see
Lois Palken Rudnick, *Mabel Dodge Luhan: New Woman, New Worlds* (Albuquer-
que: University of New Mexico Press, 1984).

the fence. Close to the river someone has severed one strand with a wire cutter.

Bushes grow in a tangle at both ends of the old wooden bridge. Only a few globs of faded asphalt still coat the crossbeams. Whole planks are rotted through, and near the middle, five-foot gaps make it frightening, if not impossible, to pass safely over the swirling brown river. The construction crew never replaced the small sign that had identified the bridge for tourists, and there is no trace of the house where Edith Warner once provided chocolate cake and reassuring conversation to J. Robert Oppenheimer and the other scientists who tried to make war obsolete with a scientific invention.

"My friend was wrong who said that this country was so old it does not matter what we Anglos do here," Warner wrote in her journal. "What we do anywhere matters but especially here. It matters very much. Mesas and mountains, rivers and trees, winds and rains are as sensitive to the actions and thought of humans as we are to their forces. They take into themselves what we give off and give it out again."[2]

Contemporary efforts to regain the lost world of unselfconscious innocence in a pretechnological paradise are easily ridiculed as naive and vulgar romanticism. In Western culture the vision of life in harmony with nature was superseded by the goal of mastery over nature long before Warner and Lawrence took refuge in the raw mountains of northern New Mexico. But somehow the dream of harmony persists, even in the shadow of the bomb.

Shrouded in secrecy, the world's first nuclear weapons laboratory was staffed by people who had been drawn to science by their own more modern and civilized version of a romantic motive: to uncover the order and structure of nature "for the greater glory of God and Man." Their motive for using those natural laws to help end the war was a sense of great moral urgency. The second World

---

[2]Church, *House at Otowi Bridge*, p. 18.

War, coming so soon after the war that was supposed to end all wars, compressed their sense of the future. The important thing, they thought, was that the Allies should invent the atomic bomb before the Germans or the Japanese. Almost no one asked how the world would be changed if the invention worked.

Most of the Manhattan Project scientists went back to their university laboratories after the Japanese surrendered. At Los Alamos, a new generation of weapons researchers interpreted science for the greater glory of God and man to mean technology to enhance the military and political power of the United States. The postwar scientists and engineers have since invented an arsenal powerful and diverse enough to make the pessimism of the 1920s intelligentsia look like the self-indulgent cynicism of an oversensitive adolescent. Fission devices were almost immediately supplemented by fusion weapons with theoretically unlimited destructive power. Simple air-drop bombs were elaborated into weapons that can be delivered by rockets, submarines, ships, and individual soldiers. One weapons laboratory became three, and together with private contractors, they became the scientific and engineering core of the military-industrial complex.

If Los Alamos is a metaphor for America, as one woman said, then it, and the other weapons laboratories, are also a metaphor for the science-centered world. Imagine that the Land of Enchantment is really the center of the earth. The weapons laboratories are the economic and intellectual machines at the center of the center. Outside are people, mostly poor and powerless, who see the land as their mother and the sky as their father. The weapons laboratories bring them some economic benefits and the promise of a modern lifestyle; in exchange, they respect the laboratories' demands for privacy and money.

Inside the laboratories, people study and manipulate light, strange metals, and esoteric numbers, continually reinventing and refining devices designed to end the world. They do it secretly, using carefully developed intellectual skills most valued by Western civilizations: reason, objectivity, and imagination.

The logic of deterrence is astoundingly simple. The superpowers' nuclear arsenals are a doomsday machine. Once set in motion, it ruins everything we and our adversaries most cherish. Changing political relationships among the superpowers may seem reassuring, but they have few significant practical effects as long as both countries feel safe only because they agree that nuclear devastation is the ultimate sanction for aggressive behavior.

Because it depends on mutual fear, the simple logic of deterrence removes the possibility of rational choice. Arguments for the necessity of more weapons can never be refuted by that logic, as almost every new investigation of a new technique for terrifying and deterring the enemy seems both necessary and inevitable. Reason is confounded when a logical problem permits only one solution, when mutual assured destruction is both premise and conclusion.

When reasoning about the necessity of keeping up or winning the arms race starts to look more and more like an exercise in rationalization, workers in the weapons laboratories begin to wonder if they have lost track of their moral values and suspect themselves of hypocrisy. When they try to be objective about the political and moral significance of their work, they become confused. Scientific objectivity is useful for dealing with small problems, but it becomes problematic when the big moral issues are at stake. How can you be objective about analyzing a problem that refuses to be contained? If every fact is a potential key to understanding the moral significance of your own life, every fact is equally capable of serving as a defense against guilt and anxiety.

Since logic reaches a dead end with mutual assured destruction, and empirical arguments flounder in a wealth of tenuous details, weapons lab workers find it easy to dismiss their own reservations about their work. Like their predecessors, modern weapons scientists and engineers prefer to see themselves as wizards who reveal and master nature's secrets. Day to day, they worry more about how to accommodate themselves to their institutions, and wonder aloud why no one from the laboratories has ever won a Nobel Prize.

Somehow the machine that powers the center of the earth stifles creativity and imagination. With enough income to live any way they choose, most weapons lab workers in New Mexico choose to live like conformists, with furniture ordered from the Sears catalog and political ideas derived from cold war propaganda or gleaned from *Time* magazine. When they contemplate the future, their imaginations become strangely distorted. Some cannot visualize the world after a nuclear war. Some see orderly trials of weapons scientists in which they imagine they will accept a guilty verdict with humility and grace. Others place their faith in the doomsday machine. They believe we can prepare indefinitely for both genocide and suicide without ever giving in to the temptation to try it.

In the machine at the center of the world, preparing for the worst possible earthly disaster becomes a routine. Tom Grissom's simple personal solution seems hopelessly idealistic to people who feel themselves enchanted by and trapped within a huge mechanism that promises to function smoothly with or without them. The essence of their secrets is known to all, just as all know that there is no way to get away from the bomb. Harmony is defined by the machine's efficient hum, and the sacred world is lost.

RECOMMENDED READING

Arms Control and the Arms Race: Readings from "Scientific American."
New York: W. H. Freeman, 1985.

Ball, George W. "The War for Star Wars." *New York Review of Books*
(April 11, 1985), 38.

Broad, William. *The Star Warriors: A Penetrating Look into the Lives
of the Young Scientists behind Our Space-Age Weaponry.* New York:
Simon & Schuster, 1985.

Carrol, Daniel E., et al. *Nuclear Weapons and Morality: A View from
Los Alamos.* Los Alamos: Immaculate Heart of Mary Catholic
Church, 1983.

Church, Peggy Pond. *The House at Otowi Bridge: The Story of Edith
Warner and Los Alamos.* Albuquerque: University of New Mexico
Press, 1960.

Cohn, Carol. "Slick 'ems, Glick 'ems, Christmas Trees, and Cookie Cutters: Nuclear Language and How We Learned to Love the Bomb."
Bulletin of the Atomic Scientists (June 1987), 17–24.

Dyson, Freeman. *Weapons and Hope.* New York: Harper and Row,
1984.

Erlich, Paul R., et al. *The Cold and the Dark: The World After Nuclear
War.* New York: W. W. Norton, 1984.

Everett, Melissa. *Breaking Ranks*. Santa Cruz: New Society Publishers, 1988.

Glasstone, Samuel, ed. *The Effects of Nuclear Weapons*. rev. ed. Washington: U.S. Atomic Energy Commission, 1962, with changes as of 1964.

Gornick, Vivian. "Town Without Pity." *Mother Jones* (Aug./Sept. 1985), 14.

Gray, J. Glenn. *The Warriors: Reflections on Men in Battle*. New York: Harper & Row, 1967.

Holmes, Jack E. *Science Town in the Politics of New Mexico*. Albuquerque: Division of Government Research, no. 71, University of New Mexico, 1967.

Jungk, Robert. *Brighter than a Thousand Suns*. New York: Harcourt, Brace, 1958.

Kahn, Herman. *On Thermonuclear War*. Princeton: Princeton University Press, 1961.

Kane, Joseph. "Los Alamos: A City upon a Hill." *Time* (Dec. 10, 1979).

Kevles, Daniel J. *The Physicists*. New York: Vintage, 1979.

Kunetka, James W. *City of Fire*. Albuquerque: University of New Mexico Press, 1979.

Lifton, Robert Jay. *Death in Life: Survivors of Hiroshima*. New York: Random House, 1967.

Lifton, Robert Jay, and Richard Falk. *Indefensible Weapons: The Political and Psychological Case Against Nuclearism*. New York: Basic Books, 1982.

Loeb, Paul. *Nuclear Culture: Living and Working in the World's Largest Atomic Complex*. New York: Coward, McCann, & Geohegan, Inc., 1982.

Los Alamos: Beginning of an Era 1943–1945. Los Alamos: Los Alamos Scientific Laboratory, no date.

"Los Alamos Science": Fortieth Anniversary Issue. vol. 4, no. 7 (Winter/Spring 1983).

Masters, Dexter. *The Accident*. New York: Knopf, 1965.

McPhee, John. *The Curve of Binding Energy*. New York: Random House, Ballantine Books, 1975.

Mojtabai, A. G. *Blessed Assurance: At Home with the Bomb in Amarillo, Texas*. Boston: Houghton Mifflin, 1986.

Morland, Howard. *The Secret that Exploded*. New York: Random House, 1981.

National Conference of Catholic Bishops. *The Challenge of Peace: God's Promise and Our Response*. Washington: Office of Publishing Services, no. 863, United States Catholic Conference, 1983.

Powers, Thomas. "Choosing a Strategy for World War III," *Atlantic* (November 1982) 82–110.

Recommended Reading

Rhodes, Richard. *The Making of the Atomic Bomb*. New York: Simon & Schuster, 1986.

Salholz, Eloise, et al. "An Identity Crisis at the 'Mesa of Doom'". *Newsweek* (October 31, 1988).

Scheer, Robert. *With Enough Shovels: Reagan, Bush, and Nuclear War*. New York: Random House, 1982.

Schell, Jonathan. *The Fate of the Earth*. New York: Knopf, 1982.

Smith, Alice Kimball. *A Peril and a Hope*. Cambridge: MIT Press, 1965.

Smoke, Richard. *National Security and the Nuclear Dilemma: An Introduction to the American Experience*. Reading, MA: Addison-Wesley Publishing Co., 1984.

Szasz, Ferenc Morton. *The Day the Sun Rose Twice: The Story of the Trinity Site Nuclear Explosion July 16, 1945*. Albuquerque: University of New Mexico Press, 1984.

Walzer, Michael. *Just and Unjust Wars: A Moral Argument with Historical Illustrations*. New York: Basic Books, 1977.

Weart, Spencer R. *Nuclear Fear: A History of Images*. Cambridge: Harvard University Press, 1988.

INDEX

and the sanctuary movement,
165–66
and security restrictions, personal responses to, 107–11,
115–16
status of, 15–16, 66–67
and stress, 55–56, 58, 94
surveillance of, 16
transfer of, 56. *See also* Manzano
Base
Schell, Jonathan, 186
Science
ambivalent power of, 34–35,
70–77

Science (*continued*)
—centered world, 228–30
constraints on, Manley on, 64
and nature, predictability, 174
and objectivity, 34–35, 38, 39,
69, 228
and transcendence, 38
and truth, 64, 66, 68–69, 175.
See also Scientific method;
Scientific mind
Science, 35
Scientific method, 5, 67
Scientific mind, 66, 67–68
and common sense, 73
and emotional sensitivity, 88
and nuclear dreams, 86–88
numerical precision of, 51
and objectivity, 34–35, 38,
87–88. *See also* Objectivity
and religion, 37–38
and subjectivity, 38
view of weapons as technical devices, 83
Security forces, 11–12, 13, 23,
109–10
Security measures, 26, 102–16,
131, 165
and fear of discussing moral and
social issues, 5, 177

and liberal staff members,
182, 183
and reshuffling of work groups,
17
Seismic monitoring, 14
Shulman, Howard, 118–119
Skinner, B.F., 220
Solar power, 9, 13, 18, 67, 70. *See
also* Energy, alternative
South Korea, 112
Soviet Union, 4, 93–94, 188, 205,
214
and espionage, 17, 111, 112
nuclear attacks by, potential, 23,
40–41
nuclear targets in, 152
staff members' views of, 125–26,
129, 131–32
weapons tests by, 53, 65
—arms race with, 69, 85, 146–
47, 186, 204
rationalization of, 229
and SDI, 192. *See also* U.S.-So-
viet relations
Spying, 17, 111–14
Stalin, Joseph, 132
Star Wars (Strategic Defense Initiative), 18, 65, 66, 89,
98, 148
debate over, 99–100, 186–91,
193–94
under Reagan, 18, 66, 110
research, in the X-division at Los
Alamos, 188, 204
scandal, 188–91
SDIO planners and, 188–91
support for, 155, 209
Teller's position on, 180
Suicide, 39–40, 69, 97, 230
Superconducting Supercollider, 200
Szilard, Leo, 142–43, 179

Teller, Edward, 40, 143
and H-bomb research, 53